TRIUMPH
TR

TRIUMPH
TR

James Taylor

Windrow & Greene

This edition published in Great Britain by
Windrow & Greene Ltd,
5 Gerrard Street,
London W1V 7LJ.

© Windrow & Greene 1997

Produced for Windrow & Greene
by The Shadetree Press, East Sussex.

Printed in Hong Kong through
Bookbuilders Ltd.

A CIP catalogue record for this book is available
from the British Library.

ISBN 1 85915 040 3

Page 3
*For the TR4, TR4A, TR5 (pictured) and TR250,
Triumph used a new Michelotti designed body
with sharper, more up to date styling.*

Acknowledgments

The author and publishers would like to thank the following for their immense help in the preparation of this book: David Bradbury (owner of the blue TR4A), Dave Destler of *British Car* magazine (who provided pictures of the TR250 and other US-spec. cars), Mark Dixon (who photographed one of the TR3s), Jerry at Enginuity (who located the blue TR6 and others), Geoff Mansfield of the Northern TR Centre (owner of the white TR5 and temporary owner of the green TR2, red TR4 and other cars of which detail pictures appear), Brian Medhurst (owner of the green TR3A), Bill and Graham Rimmer of Rimmer Bros (owners of two TR8s), Alan Thomas (owner of the gold TR7), and to the many TR enthusiasts who never knew that their cars had been photographed! Additional photographs were provided by Mike Key and the TR Drivers Club. Finally, thanks to Haymarket Publishing, publishers of *Autocar*, for allowing the use of various TR road-test articles.

Contents

Introduction

Mention the TR name in motoring circles today, and you will probably find that it conjures up visions of rapid, affordable, two-seater sports cars, which were always rather crude in conception and usually simple to work on. Even though many of the last generation of TRs were relatively refined closed cars, it is the image created by the open TRs that lingers on.

In nearly 30 years of production, there were several different models of Triumph TR. The original range gradually evolved from the TR2 into the TR6 over a decade and a half, always on the same wheelbase, but with progressive increases in engine size and power output. On the way, the range picked up independent rear suspension (with the TR4A) and fuel injection (with the TR5). It switched from the in-house styling of the TR2, TR3 and TR3A models to the Michelotti styling of the TR4, TR4A and TR5, and thence to the Karmann restyle of the TR6. The TR7 represented the first break with this long tradition, a tradition so well established that even now, more than ten years after the last TR7 was built, many TR enthusiasts still refuse to accept these cars as 'real' TRs.

The most significant factor in the TR story from start to finish was the US market. The cars were designed for sale in North America from the beginning, and the vast majority of those built were sent across the Atlantic. The first special variant designed to suit North American tastes was the TR3B of 1962, but by the

Facing page, top
The first production Triumph TR sports car was the TR2, introduced in 1953.

Facing page, bottom
The TR3 continued the theme from 1955. Sales rose by more than 50 per cent over the TR2, while export sales more than doubled.

Left
The last of the 'sidescreen' TRs was the TR3A, introduced in 1957.

end of the decade, US emissions control regulations ensured that TRs for that market were markedly different from the home-market cars. In Britain, we tended to think of the US cars as variants of the basic models, and indeed Triumph's model nomenclature reinforced that illusion. But, in terms of numbers at least, the US models were the real mainstream cars, while the British examples were the low-volume special variants.

By the early 1970s, however, emissions controls had sapped the performance of the North American TRs, build quality problems had sapped customer confidence, and the arrival of more refined machinery, like Datsun's 240Z, had caused Triumph to struggle in the US

market. The TR7, then, was an all-new design intended to recapture a lost market. So specifically was it aimed at the US buyer that it did not even go on sale in Britain until 12 months after it had been introduced in the US. And its more powerful TR8 derivative never went on sale anywhere except in the USA.

Unfortunately, the writing was already on the wall by the time the TR8 arrived. Constant trouble with a demoralized work-force had led to serious delivery delays and to ever worsening quality control. Things did begin to improve, but the damage to the TRs' reputation had already been done. Triumph decided to cut its losses, and the last of 373,430 TRs was built in October 1981.

Right
For 1961, the TR4 featured restyled bodywork by Michelotti. This example is the rare GTR4 coupé built for Triumph dealers L. F. Dove.

Below
Minor changes were made to Michelotti's styling for the TR4A, which featured an uprated chassis to match the more modern looks.

Left

The first of the six-cylinder cars, the TR5, appeared in 1967, but its fuel injection system did not meet US emissions standards, so a special carbureted model, the TR250, was developed.

Left

Last of the old-style, separate-chassis TRs was the TR6, which featured styling by Karmann and outsold all previous TRs by a considerable margin.

Below

The monocoque TR7 made a radical break with TR tradition. This is one of the first cars for the UK-market, built in 1976.

Above

The final TR was the TR8, which was sold only in the USA, although some right-hand-drive versions were built for evaluation in the UK.

THE TRIUMPH TRs—A CHRONOLOGY

1952 (October)	Prototype TR1 shown at Earls Court motor show
1953 (March)	TR2 prototype shown at Geneva motor show
1953 (August)	TR2 production begins
1955 (October)	TR2 production ends
	TR3 production begins
1957 (September)	TR3 production ends
	TR3A production begins
1961 (August)	TR4 production begins
1961 (October)	TR3A production ends
1962 (March)	TR3B production begins
1962 (October)	TR3B production ends
1965 (January)	TR4 production ends
	TR4A production begins
1967 (August)	TR4A production ends
	TR250 production begins
1967 (October)	TR5 production begins
1968 (November)	TR5 production ends
	TR6 production begins
1968 (December)	TR250 production ends
1975 (January)	TR7 production begins
1976 (July)	TR6 production ends
1979 (September)	TR8 production begins
1980 (March)	TR7 and TR8 convertible production begins
1981 (October)	TR7 and TR8 production ends

By model-year				
	1954-55	TR2		
	1956-57	TR3		
	1958-61	TR3A		
	1962-65	TR4	1962	TR3B (USA)
	1965-67	TR4A		
	1968	TR5	1968-69	TR250 (USA)
	1969-76	TR6		
	1975-81	TR7	1980-82	TR8 (USA)

The first TR

If it had not been for the success of the MG TD and Jaguar XK120 in the USA, there would never have been a TR sports car from Triumph. However, Standard-Triumph's chairman, Sir John Black, was envious of the sales that these two sports models had achieved across the Atlantic, and determined that his own company would cash in on the apparently insatiable demand for sports two-seaters.

Standard-Triumph's first attempt at fulfilling Sir John's dream was based on the Standard Vanguard chassis and running gear, and was a rather bulbous-looking creation known as the TRX. It was going to be expensive to make, and its most sporting feature was its convertible top. By the end of 1950, Sir John had been persuaded not to proceed with it.

As the TRX project died on its feet, Sir John considered an alternative plan: Standard-Triumph was already supplying engines to Morgan for use in that company's sports cars; why should it not buy the company and use its products to attack the US sports car market? Unfortunately for Sir John, the Morgan management had other ideas and declined his offer of a take-over in December 1950. The only solution left, therefore, seemed to be for Standard-Triumph to have another go at developing its own sports car.

The company's design engineers were far too busy with other commitments during 1951 to find any time to devote to a new sports car but, at the beginning of 1952, the project got under way. Sir John Black instructed his designers to come up with a sports car that used as many existing Standard-Triumph components as possible, and told them that they could

have what now appears to be a derisory sum to develop a new body for it.

To save money, the designers initially proposed to base the car on the pre-war Standard Flying Nine chassis (of which a large surplus stock existed), but by the time of the first prototype so many alterations had been made that this was no longer feasible. They added independent front suspension from the current Triumph Mayflower saloon, a narrow-track version of the Mayflower's rear axle, and a Standard Vanguard engine, fitted with cylinder liners to produce a capacity of less than 2 litres and thus make it eligible for club sporting events. The gearbox, too, came from the Vanguard, although it was redeveloped from a three-speed into a four-speed type. The whole was topped off by a very simple two-seater body, in the modern, all-enveloping idiom popularized by the XK120.

The single prototype was finished just in time to be shown at the 1952 Earls Court Motor Show. Even before it went on display, the designers knew that certain elements needed further work. Nevertheless, reactions to the car—which was simply referred to at the show as the Triumph Sports Car—were encouraging. The short tail with its exposed spare wheel mounting came in for criticism, but that could be redesigned without much difficulty.

The other question was whether Triumph still had a credible sporting pedigree, as its post-war models had so far completely failed to recapture the glories of the pre-war marque. For Sir John Black, however, this was not an issue. He was determined that his new sports car would succeed.

TR2

As 1952 drew to a close, the Triumph engineers set to work on their sports car prototype—known within the factory as the 20TS—to get it ready for production. Redesigning the rear end of the body was simple. However, Sir John Black also wanted to make sure that performance and handling were up to scratch, so he asked Ken Richardson, formerly involved with the BRM Grand Prix car, to test-drive the Triumph prototype. Richardson expressed some serious reservations—and was promptly offered a job with Triumph as project engineer to help effect the necessary changes!

The main alterations that Richardson recommended were to the chassis, which was made stiffer. In addition, some development work was carried out on the engine to increase its power output to 90bhp (67kW) from the 75bhp (56kW) of the first prototype. By the beginning of 1953, all the redesign work had been completed, and the first prototypes of what was now being called the TR2 were undergoing endurance testing. The public had its first chance to see the redeveloped car at the Geneva show that March.

Keen to gain some publicity for the new car, Sir John Black arranged for one of the prototypes to carry out timed high-speed runs on the Jabbeke motorway in Belgium. No doubt he had been inspired by the example of Rootes, which had made similar runs to announce the new

Facing page
Proudly displayed on the bonnet of this car is the enamelled TR2 badge, almost the only identification that the car carried.

Below
The TR2 was a stylish two-seater that offered value for money in terms of performance and economy. It was intended to capitalize on the success of the MG TD and Jaguar XK120 in the USA.

Sunbeam-Talbot sports model. So it was that, on 20 May 1953, Ken Richardson drove one of the TR2 prototypes at speeds approaching 125mph (201km/h) on the Jabbeke motorway. On the same occasion, the car demonstrated that it was capable of 109mph (175km/h) even in touring trim. This was exactly what Sir John Black wanted to give the TR2 the right sort of image when it finally became available that autumn.

In performance and in price, the TR2 was pitched between the MG sports model (by now the TF, although still very traditional in design) and the Jaguar XK120, which immediately gave it a niche all of its own. Public and press responses were both positive, but sales were slow in the beginning because production took time to get under way. Sales in the USA were certainly not as great as Triumph had hoped for; in fact, exports pretty well collapsed during 1955, thus

Above
In addition to the bonnet badge, the only other means of identification were the Triumph 'globe' badges on the TR2's hub-caps. Very early examples did not even have these, but had plain centres to their hub-caps. This car has silver wheels, but originally they would have been body colour.

Right
The TR2's lines looked good from behind. Note the large and prominent fuel filler cap just ahead of the boot lid: this was intended to add to the car's sporting image by evoking the wide-necked fillers used in motor racing to facilitate rapid refuelling during pit-stops. The boot lid was held shut by a 'budget' lock at each side, which was opened by a T-handle key supplied with the car. The locks were covered by chromed escutcheons.

releasing large numbers of TR2s on to the home market.

Meanwhile, the cars had been racking up competition successes in private hands. At the Earls Court show in October 1954, a number of refinements had also appeared: better brakes, an optional hard-top and shorter doors (the originals tended to catch on high kerbs). But for 1956, Alick Dick, who had replaced Sir John Black early in 1954, was determined that the TR would do better. As a result, the TR2 remained in production for only two years before it was replaced.

The TR2 spawned only one rarity. This was called the Francorchamps and was converted from a standard, factory supplied car by Imperia in Belgium, who were Triumph's importers. The Francorchamps had a fixed hardtop with a clear plastic sun-roof and rear window, and other special features. Only 22 were built during 1954-55.

Above
The 1991cc four-cylinder engine of the early TR2s pumped out 90bhp (67kW). This example has been restored superbly. The windscreen washer bottle was not fitted when the car was new, but was made necessary by subsequent legislation.

THE TESTERS' VIEW

Post-war production of the Triumph Division of the Standard Motor Company consisted initially of two versions of the Triumph 1800, a knife-edge saloon and a roadster, both cars being able to seat up to five, although in the roadster two persons were carried in occasional seats in the luggage locker. Later, 2-litre power units, similar to the Standard Vanguard engine, were fitted to both these models. The roadster was a car of sporting character with many of the refinements associated with a drophead coupé and not an out-and-out sports car like the latest model to be produced by the company, first introduced at the 1952 London Show, but only lately coming into production, which is the subject of this test.

In the main, the type of person to whom a sports car appeals is the enthusiastic and often youthful driver. He wants a car with performance, yet he often has very little money to pay for it. There are a number of models that would perhaps fully meet his requirements if only he could afford the price, while those cars that he can afford do not have the performance that he desires. The sports Triumph not only provides an outstanding performance, but it is also particularly good value for money as regards both initial purchase price and running costs, as a glance at the fuel consumption figures will show. Bearing in mind that a figure of 124 m.p.h. *(200km/h)* for the flying mile was obtained with a car in speed trim in demonstration runs on the Jabbeke road early last summer, this journal eagerly awaited the opportunity of trying out a car in standard trim to measure the performance obtainable without the use of items such as an undershield and metal cockpit cover.

Results were in no way disappointing. From a flying start the car tested attained its maximum speed over a test distance of two miles *(3km)* with hood up and sidescreens in position. Over this distance no increase in maximum speed was obtained by using the overdrive, although it is possible that the absolute maximum on overdrive might be higher given an unlimited stretch of straight level road. The car was also tested with the hood down and sidescreens removed but with the normal windscreen in position, and in this trim a mean speed of 99 m.p.h. *(159km/h)* was

obtained, showing, as would be expected, that the car is slightly slower when open unless the windscreen is removed and a full tonneau cover fitted.

There are very few cars indeed that have a mean maximum speed of well over 100 miles an hour *(161km/h)* in standard trim and at the same time sell for about £600 basic price, and register an overall fuel consumption of 32 m.p.g. *(11km/litre)*. From these few items of performance, it might be thought that this is one of those cars where everything has been sacrificed in the interests of performance, but this is not so, for good as this is, the TR2 is by no means stark. In fact, it is very well finished and equipped, and creates the impression even

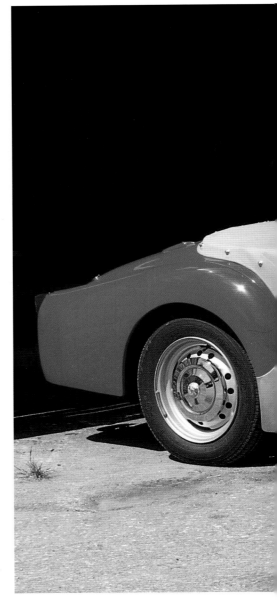

Left
The TR2's spare wheel lived behind its own access panel in the rear, and its position explains why there was no proper rear bumper. The central red light, above the number-plate, was actually a brake light. The wing mounted lights served as tail-lamps and as flashing indicators.

Below
A neat tonneau cover provides protection for the cockpit when going topless.

Below

The hood frame was a very simple tubular affair that folded flat behind the seats; the hood itself lived in the boot. Although it was easy to erect in theory, in practice the driver and passenger could become very wet before they achieved a satisfactory fit!

Bottom

The dashboard was simple, but the instruments were neatly laid out and clear to use—as they would be on all the TRs. This car is fitted with overdrive, controlled by the knob just visible above the wheel spokes on the right of the dashboard.

after only a brief acquaintance of being a well-balanced car that has that satisfying "all in one piece" feeling, an impression that grew as experience with the model increased.

Based on the well-known Vanguard engine, the power unit for the Triumph is of a slightly smaller capacity to bring it within the 2-litre class. It has a higher compression ratio, a different camshaft with modified valve gear, and is fed by two side draught S.U. carburettors. In this form it develops over 45 b.h.p. *(33kW)* per litre, so that when fitted in this light two-seater it has 81 b.h.p *(60kW)* per ton laden. The engine is smooth and has a satisfactory degree of silence, although there is a little valve noise—the silent valve gear used in the Standard Vanguard is not employed for this sports model. On first grade fuel some slight pinking is noticed when the engine is pulling hard, but this is not excessive. There is no noticeable flat spot in the carburation, and the

acceleration generally is very clean; the unit responds very well to the throttle.

Overdrive

The model tested had an optional extra in the form of the Laycock-de Normanville overdrive unit which is applied to top gear only (the ratios are such that there would be little advantage in using the overdrive on third gear); overdrive engagement is effected electrically. The clutch is both smooth in operation and well able to stand up to fast gear changes without undue slip. It has an hydraulically operated withdrawal mechanism. The pedal is comfortably light and at the same time does not have excessive travel.

In keeping with the character of the car, a central remote control gear lever is used; it is very well positioned in relation to the steering wheel and has a particularly easy and positive movement from gear to gear. It is a robust mechanism, well able to cope with full-throttle changes. The synchromesh provided on top, third and second gears of the four-speed box is effective and not easily beaten when fast changes are made. The overdrive switch is conveniently placed on the facia, where it can be operated by the driver's right hand. The change up into overdrive is quite smooth, but a slight jerk is noticed when a change down to direct top gear is made, unless the clutch pedal is lightly operated.

One of the outstandingly important things about a sports car is the way it handles. There is little point in having a high power output and low weight if the car does not behave well; in this direction the Triumph earns full marks. It has a nicely balanced feel which quickly inspires confidence. The suspension is sufficiently soft to provide a comfortable ride, yet it does not permit excessive body movement, and there is noticeably little roll on corners.

The car tested was fitted with Dunlop Road Speed tyres and it is recommended that the standard pressures of 22 front and 24 rear *(1.5 and 1.7 bar)* should be increased for high speed work; consequently, pressures of 30 lb per sq in *(2.1 bar)* front and 36 lb *(2.5 bar)* rear were used for the high speed runs. For the acceleration figures the tyres were set at 26 front, 30 rear *(1.8 and 2.1 bar)*, and the car was also driven for some distance on the road with the tyres set at this pressure. Compared with the standard settings, this increase produced a slightly harder ride and at the same

Left
For the car's size, the cockpit was surprisingly roomy, while the well-contoured bucket seats offered a good range of adjustment. The arrangement of the steering wheel spokes provided a clear view of the instruments.

Below
The positions of the headlamps, bonnet badge and radiator opening gave the car a distinctive 'face'. For competition, the frontal area could be reduced by removing the windscreen.

FCH 997

Right

This car bears the factory-issue optional badge bar, to which is mounted the badge of the original Triumph Sports Owners Association, formed by the factory itself in 1954. Years of battering by road dirt and debris have removed almost all of its red and black enamel finish.

Below

This TR2 is a 1954 model, with the original 'long-door' bodywork. Like so many early TRs, it was finished in British Racing Green and, like so many others of its age, it has been modified over the years. When new, the wing beading would have been in body colour rather than chrome.

time made the steering feel a little lighter.

The general layout of the car results in a slight amount of understeer which further increases the general directional stability. Roadholding on corners is particularly good, no matter whether they be fast, open bends or sharp curves. Roadholding is also very good on wet surfaces. In keeping, the steering is very positive: there is an ample lock, and $2\frac{1}{4}$ turns of the steering wheel from lock to lock enable a quick change of direction to be made without excessive wheel turning. The system is light, yet it has an accurate feel so that the driver knows that he is in contact with the road, yet no appreciable road shock is transmitted back through the steering wheel.

The hydraulically operated brakes have two-leading shoes at the front; they are very powerful, and under test conditions recorded a particularly good efficiency. Under the strenuous conditions imposed during the actual performance testing, where repeated brake applications occur at very frequent intervals, the rise in brake temperature made it necessary to

apply increased pedal pressure, but there was no suspicion of loss of balance, and, if the brakes were applied hard, perfectly straight black lines could be produced on the road surface. The hand brake lever, too, is effective; it is mechanically coupled to the rear wheels and fitted with a fly-off type of ratchet. Over the test mileage there was no noticeable increase in pedal travel, although there was a slight tendency for the brakes to squeak at times with light pedal pressure.

Apart from a healthy but not unpleasing bark from the exhaust over a limited speed range around 2,400 r.p.m., the car is generally very quiet. There is a little engine noise but the rest of the mechanical components are quiet. There is very little wind noise and the weather equipment does not flap from the effects of the wind. The car is also free from vibration.

In order to obtain the best results from the horse power available it is necessary to reduce the frontal area of the car as much as possible, and, if this is carried to the extreme, it might adversely affect the general passenger comfort. In the Triumph, in spite of the front area of $15\frac{1}{2}$ sq ft *(1.4sq m)* with the hood up, the passenger and driving compartment is not unduly cramped; in fact, there is a surprising amount of space. Driving comfort is important in any vehicle but it is particularly important in a sports car, especially if it is to be used for competition purposes—no one would expect a cricketer to perform well if he were given a bat two sizes too small for him! In spite of the compact overall dimensions of the Triumph, drivers of most sizes can be comfortably accommodated in it.

Both seats have a useful range of adjustment, and the relationship between steering wheel and pedals is very good. The seats themselves are well upholstered and give good support, although it would be better if there were a little more clearance between the driving and passenger seats. If the passenger seat is placed closer to the toe-board than the driving seat, the back rest tends to get in the way of the driver's left arm, but this matter is being attended to on future production cars. The relationship between the brake and throttle pedals enables heel and toe gear changing methods to be adopted, and the dip switch provides a rest for the driver's left foot. The hand brake lever is placed on the right-hand side of the central tunnel which encloses the

gear box; in consequence it protrudes into the driving compartment and tends to get in the way of the left leg of a big man. It is very well placed for convenient operation.

With the hood erected there is a satisfactory amount of head room even for a fairly tall driver, and from the driving seat the forward visibility is very good and the head lamp cowls and both front wings can be seen. The windscreen pillars are very slender and do not tend to cause a blind spot. With the side screens erected the all round visibility is quite good, but it would be better if the rear window area were increased to cut down a blind spot caused by the rear quarters of the hood. The mirror is well placed so that it does not mask a large area of the screen, and at the same time it provides a satisfactory rear view.

To prevent reflections in the windscreen, the steering wheel has a black rim and the arrangement of the T spokes permits a clear view of the speedometer and tachometer mounted in the facia panel in front of the dri-

Above
Only the early TR2s had this kind of bonnet locking mechanism, operated from inside the car. On 1955 models, the bonnet was secured by twin Dzus fasteners, operated by a turnkey from outside.

ver. Other instruments, which include an ammeter, fuel, and water temperature and oil pressure gauges, are mounted in the centre of the facia together with the control switches. All the instruments are effectively illuminated, and the position of the facia, which is set back underneath the scuttle, reduces the reflection caused in the windscreen at night. However, some reflection is caused by the bottoms of the dials placed in front of the driver. Twin wipers pivoted at the bottom of the screen cover a wide area of the glass, and the arcs of the blades overlap so that the whole of the central portion of the windscreen is kept completely cleaned.

Both doors are well fitting and free from rattles, but the bottoms of the doors do not have sufficient clearance to enable them to be opened when the car is parked close to the kerb. The interior of the car is well trimmed and nicely finished. Pockets are provided in both doors and there is a lockable compartment on the passenger side of the facia panel.

Above
The chassis number of a TR2 was carried on this plate, screwed to the engine side of the front bulkhead.

Left
This side view shows clearly the rakish, slightly aggressive lines, that would characterize all TRs until Triumph introduced the Michelotti styled TR4 in 1961. The stone-guards on the leading edges of the rear wings were an attractive styling feature, as well as being functional. The hood frame, seen here exposed, was provided with a neat cover in material that matched the upholstery. A patriotic owner has fitted small Union Jack badges to the bonnet sides of this car, but these did not form part of the original specification.

Above
The 1954-model TR2s were the only TRs to have full-length doors. These tended to catch on low kerbs when open, so 1955 models had shorter doors with visible outer sill panels below them.

Facing page
This picture shows the sill panel that was normally concealed by the door. The door trims incorporated pockets, and there were door pulls instead of handles. The chromed sockets accommodated the support sticks of the sidescreens. The relay visible below the facia is for the optional overdrive fitted to this car.

Another useful fitting is a grab handle placed just above this locker. The hood is made of plastic material, it fits well and is easily dismantled and erected. With all the weather equipment in position the car does not leak water around the joints of the side screens and hood, but there is a certain amount of draught when it is driven fast; this is not excessive and could easily be offset by the addition of normal heating equipment, which is available as an optional extra.

Apart from the rear luggage locker there is also a useful space behind the seats which could accommodate a suitcase. The main locker itself is of a useful shape and of reasonable proportions, bearing in mind the size of the car. A separate compartment below the locker is used to house the spare wheel. The fuel tank, located above the rear axle, has a central spring-loaded filler cap, and can be filled quickly without blowing back. It is provided with an overflow pipe. With its capacity of 12½ gallons *(57 litres)* it gives a particularly useful cruising range.

The double dip head lamps provide a useful main beam range and give a satisfactory spread of light in the dip position. A winking type of direction indicator is used, and it would be better if the indicator lights were brighter. Starting from cold was at all times very good, and the spring-loaded choke control could be released almost immediately the engine had fired. Thirteen points on the chassis require lubrication with a grease gun at intervals of 1,000 miles *(1600 km)*. Two jacking points are provided, one on each side of the frame, and to gain access to these it is necessary to remove rubber bungs from the floor.

The sports Triumph is particularly good value for money. It has a fine performance and it is also very economical on fuel. Added to these qualities, it is fun to drive.

Reprinted from The Autocar, 8 January 1954.

TRIUMPH TR2—SPECIFICATIONS

Engine	Four-cylinder, OHV, carbureted
Capacity	1991cc (83mm bore, 92mm stroke)
Max. power	90bhp (67kW) @ 4800rpm
Max. torque	117lb/ft (174kg/m) @ 3000rpm
Transmission	Four-speed manual gearbox with optional overdrive
Suspension, front	Independent, with coil springs, wishbones and telescopic dampers
Suspension, rear	Live axle, with semi-elliptic leaf springs and lever-arm dampers
Steering	Cam-and-lever
Brakes	Drums on all four wheels
Tyres	5.50x15 crossply
Length	12ft 7in (3.84m)
Width	4ft 7.5in (1.41m)
Height	4ft 2in (1.27m) (with soft top erected)
Wheelbase	7ft 4in (2.24m)
Max. speed	103mph (166km/h) (with overdrive)
0-60mph (97km/h)	11.9sec
O'all fuel consumption	32mpg (11km/litre)
Production total	8628 (2823 for home market; 5805 for export)

TR3

The TR3 was not a radically differ-ent car from the TR2; rather it had evolved logically from what had gone before. The styling had been tidied up, while its power and performance had been improved. Otherwise, it was the same combination as before: open-air motoring with few creature comforts and the rugged four-cylinder wet-liner engine providing the excitement.

That this formula was hugely successful was proved both by the TR3's competi-tion successes and by vastly increased sales. The works team carried off some memorable victories—notably on the 1956 Alpine Rally—which helped to establish the TR pedigree and also did their bit for sales. Exports were booming, with the result that the home market was

almost starved of cars, even though Triumph turned out TR3s at twice the rate at which the company had built TR2s.

Like its predecessor, the TR3 remained in production for only two years. Styling changes made it easy to distinguish from the TR2: there was a new 'egg-box' grille, now at the mouth of the air intake rather than recessed, and chromed beading was fitted between the body and wings. Many cars were also fitted with the occasional rear seat option, which was little more than a joke, but may have persuaded buy-ers with young children that they could cope with a TR after all.

The most important changes applied to the TR3 were mechanical. From the beginning, the engine was more powerful than that in the TR2. It had enlarged inlet

Facing page
The TR3 is a relatively rare car in Britain, where only 1286 examples were sold in two seasons of production. In the same period, however, over 12,000 went for export. In the home market, British Racing Green remained one of the most popular colours. This 1956 model has the optional wire wheels. The two large fog/spot lamps were also optional accessories.

Left
The dashboard of the TR3 had barely changed from that fitted to the TR2. As before, there were two large dials ahead of the driver, and four circular dials grouped in the middle of the facia.

ports and bigger carburettors, which together added an advertised 5bhp (4kW) to the power output. It was also possible to buy the car with a 4.1:1 axle ratio, which improved its sprinting ability at the expense of top-end performance. However, the original TR3 specification would not last long.

During the first half of 1956, Triumph introduced engine changes. First of all came the so-called 'Le Mans' cylinder head, based on that used for the 1955 Le Mans works team cars; then came the 'high port' head, which combined the best elements of the 'Le Mans' head with those of the original TR3 head. Both offered a further 5bhp (4kW) over the late-1955 TR3 engine. For most of the first half of 1956, cars were produced with both types, more or less at random.

For 1957, however, the TR3 was changed quite significantly. The 'high port' head was standardized, and a stronger axle—borrowed from the Standard Vanguard Phase lll—replaced the old Triumph Mayflower type. But most important was the introduction of disc brakes on the front wheels, making the 1957 TR3 the first series-production British car to be equipped with such brakes as standard.

Buyers of the 1957 TR3 could also order an optional 'GT' kit, which had been developed to allow the works team cars to compete in Grand Touring as well as Sports categories. This consisted of a steel hardtop (instead of the standard GRP type) plus exterior door handles.

Left
The simplest way of distinguishing a TR3 from a TR2 at a distance is by means of the radiator grille. On the earlier cars, it was recessed; on the TR3, it was fitted at the mouth of the air intake. This is a 1957 example, once again equipped with wire wheels and finished in British Racing Green.

Left
Long, low and sleek—at least, that's how the TR3 appears in this picture taken by Mark Dixon of the 1956 car. The wire wheels suit the lines of these early TRs superbly.

THE TESTERS' VIEW

For the TR sports models made by the Triumph company and delivered in considerable numbers to purchasers at home and abroad, the past three years have brought steady development without major redesign. The original two-seater had an excellent reception and at once began to distinguish itself in competition. A full Road Test of the the TR2 appeared in *The Autocar* of January 8, 1954, and of the hardtop model on February 18, 1955. Since then a number of changes have been made in the specification, among the most important of which is the adoption of disc brakes for the front wheels on the latest model here described.

In summarizing the specification, one thinks at once of the excellent 1,991 c.c. engine, which is smooth for such a large four-cylinder, gives plenty of power, and additionally has earned a reputation for economy and long life. In its latest form the engine, which is essentially similar to the unit of the Standard Vanguard, gives 100 b.h.p. *(75kw)* at 5,000 r.p.m., 10 b.h.p. *(7kW)* more than the output reached on the earlier cars at 4,800 r.p.m.

Despite the power increase, and the smoother body shape with hardtop as compared with the ordinary hood of the first car to be tested, some of the performance figures are not quite as good as those of the earliest car. However, they are a little higher than for the TR2 hardtop. In an economically produced car there is often some small variation in performance between one example and another, but the main reasons for the present car not showing up quite so well are that the weight is now greater and the weather conditions on the Belgian autoroute, where the acceleration testing took place, were decidedly worse than those prevailing for the earlier tests. There was a stiff diagonal breeze, which was particularly noticeable at the higher speeds.

Two effects which accompany the increased power and provision of the hardtop, disc front brakes and overdrive on the upper three gears are an increase in price, which now totals over £1,000 including £360 British purchase tax, and an increase in fuel consumption. High performance has always to be paid for and the latest car still represents good value. The latest m.p.g. figures are also creditable for such a fast car.

How does the TR3 hardtop behave on the road and in what way have the modifications affected its handling characteristics? One difference was unexpected: the stability of the back end on corners is not quite so good, particularly if the road surface is poor. The car will get from A to B remarkably quickly, but care is desirable if the tail is not to skip out occasionally, or bump-skid on second-class roads. The high-geared steering enables immediate corrections to be made, and the skittishness of the rear, therefore, is not of a kind to bring trouble.

A possible cause of the handling difference compared with the previous car is that there is now a bigger, heavier, altogether stronger rear axle assembly. Strangely enough, during the

test, the unit failed from what may fairly be called a freak fault: some of the bolts holding the crown wheel to the differential cage sheared. A replacement of the earlier type was immediately despatched to the local Belgian agent by the Standard and Triumph concessionaires for that country, and this was used for most of the test and performance measurement.

General stability is of a fairly high standard, although driving can become a little tiring on fast, fairly straight roads when there is any wind. In such conditions the driver must keep the car straight by conscious effort. A greater tendency for the car to follow its nose would then be welcome.

The car reaches 80 m.p.h. *(129km/h)* so quickly that even on busy British roads such speed may be used frequently and in safety. After this point the acceleration falls off, yet

Left

The red and black enamelled shield badge on the TR3's nose was essentially the same as that fitted to the TR2, except of course that it bore the new car's model name.

Left

A comparison of this picture with the rear view of the TR2 on page 15 shows further TR3 recognition points. Note particularly the chromed 'Triumph' nameplate on the tail panel. This 1956 car has the central stoplight that was fitted to all but the very last TR3s. In this case, the chromed wing beading and boot lid hinges are to original specification.

more than a genuine 90 m.p.h. *(145km/h)* is attainable on any longish main road run. For the high maximum of over 100 m.p.h. *(161km/h)* to be reached safely, a real motor road, of the kind found on the Continent or across the Atlantic would be desirable.

The Laycock-de Normanville electrically operated overdrive is used normally only on top gear, the first three orthodox gears being more than adequate for building up speed quickly. Overdrive was not used at all in obtaining standing start acceleration data, but on the main road, second or third gear overdrive is particularly convenient when a higher ratio is wanted quickly while overtaking other traffic. The change into overdrive is smooth when accelerating at fairly high engine speeds, but just a little jerky when re-engaging normal gear.

On the earlier models tested, overdrive operated only on top gear. The advantages of overdrive are appreciated particularly while fuel is in short supply, in a powerful light car with a small frontal area. This model may be persuaded quite easily to give a substantial economy, cruising effortlessly at quite high speed in overdrive top. In town the overdrive available on second and third may be used to smooth out traffic restrictions and to save fuel. Making an effort to use little fuel a figure of 35 m.p.g. *(12km/litre)* should be obtainable.

The earlier model, unmodified but using special driving techniques, won an economy contest at 71.02 m.p.g. *(25km/litre)*.

One of the few real innovations of recent years has been the application of disc brakes, for which reason special interest attaches to those enterprisingly fitted at the front of this model. They are of Girling design and construction, and much of the story is told by the data. The Tapley meter recorded 94 per cent efficiency with a pedal pressure of 70 lb *(32kg)*, and 72.5 per cent with a modest 25 lb *(11kg)*. These figures show that the retardation power is excellent. During testing there was no deterioration of the brakes, and never a trace of fade.

The only peculiarity of discs—certainly not confined to this car—is some squeaking, particularly when cold. The noise is of high frequency, sounding at times rather like a typical French car horn at a distance. It is not loud enough to annoy. The fly-off lever, which acts on the orthodox drum rear brakes, has its handle a little far from the driver, but is not difficult to reach. Two long-legged drivers found during the test that the lever caused discomfort by vibrating against the left leg—a fault overcome by taping some sponge rubber to the handle.

The hardtop is detachable. It has a pleasing, smooth shape and is well finished. The

Below

This red TR3 was one of the last to be built, documented as the 118th from the end of production. It underwent a complete rebuild in 1986, and is seen here on show at a special exhibition in the old Triumph factory at Canley.

mounting is strong, and although a spanner is required to take the top off, the job does not take long. Plastic sliding windows are fitted in the sidescreens. To open a door from the outside, a window must be opened first so that one's hand may reach down to one of the interior door handle straps. The mating of the forward edges of the sidescreens with the windscreen is not perfect, and as a result there is some wind noise. Any slight draught in the cockpit is offset by the effective heater. All-round visibility is good, as the rear window sweeps well forward towards the sidescreens.

The instrument layout is both tidy and comprehensive, and is well suited to the character of the car. The separate bucket seats give good lateral support, but the backrests could be improved. On a long run the centre of one's back may ache unless a small cushion is inserted where it will keep the back straight.

The rev counter and speedometer can be seen through the unobstructed upper half of the three-spoked sprung steering wheel, and most of the minor controls are laid out in a row at the foot of the facia. Gauges for oil pressure, water temperature, charging rate, and fuel level are grouped in the centre of the facia, with a lockable glove compartment to their left. There are also quite large pockets in the doors. The passenger is provided with a grab handle, albeit mounted in a position which might cause injury in the event of an accident. Elbow room is just sufficient.

Entry and exit is rarely very easy in a sports car, but on this Triumph the doors have good width, and the sides of the scuttle are well cut away so that the occupant's legs may be swung in or out without much difficulty.

Luggage space is good for a car of this type. There is a locker at the rear where soft bags may be stowed over the separate spare wheel compartment, and there is a considerable amount of room behind the seats. An occasional bench seat can be provided for children.

Both the luggage locker and the bonnet can be opened only with a special, square-ended key kept inside the car. Under-bonnet accessibility is above average. The twin S.U. carburettors, dip stick, coil, distributor, oil and water fillers are all easy to get at. The battery is mounted centrally, but nevertheless it is easy to top up with one of the patent filling bottles now available, or one may inspect three cells from each side of the car.

In its latest form the TR3 is an exciting sports car, traditional to drive, fast, flexible and with first-class brakes. The steering is quick and light, and visibility good. This latest model is as pleasant when it is closed up in the winter as it is in open form in the sunshine.

Reprinted from The Autocar, 11 January 1957.

TRIUMPH TR3—SPECIFICATIONS

Engine	Four-cylinder, OHV, carbureted
Capacity	1991cc (83mm bore, 92mm stroke)
Max. power	95bhp (71kW) @ 4800rpm; later 100bhp (75kW) @ 5000rpm
Max. torque	117lb/ft (174kg/m) @ 3000rpm
Transmission	Four-speed manual gearbox with optional overdrive
Suspension, front	Independent, with coil springs, wishbones and telescopic dampers
Suspension, rear	Live axle, with semi-elliptic leaf springs and lever-arm dampers
Steering	Cam-and-lever
Brakes	Drums on all four wheels (1956 models); discs on the front and drums on the rear (1957 models)
Tyres	5.50x15 crossply
Length	12ft 7in (3.84m)
Width	4ft 7.5in (1.41m)
Height	4ft 2in (1.27m) (with soft top erected)
Wheelbase	7ft 4in (2.24m)
Max. speed	102mph (164km/h) (with overdrive)
0-60mph (97km/h)	12.5sec
O'all fuel consumption	25mpg (8km/litre)
Production total	13,377 (1286 for home market; 12,091 for export)

TR3A

Just as the TR3 had evolved logically from the TR2, so the TR3A evolved logically from the TR3. Production began in September 1957, and the first deliveries were made to the US market, which confirms how important transatlantic sales had become to Triumph. The remainder of the world—and that included the UK home market—had to wait until January 1958 for the new model to be made available. In the meantime, potential customers continued to be offered the TR3.

Badging for the TR3A remained exactly the same as for the TR3, but there were several recognition points that distinguished the new model from the old. The TR3A had a wider grille, with a sidelamp incorporated at each end; it also had as standard the door handles offered with the TR3's optional GT kit, and both these handles and the boot handle were lockable. Clearly, the market now demanded this kind of refinement rather than extra performance, for the car's mechanical specification remained unchanged from that of the TR3.

Sales boomed. In the three years of the TR3A's production, an average of 19,000 cars were sold each year, which meant that it was selling some 50 per cent more every year than the TR3 had sold for the duration of its two-year production run. Most cars, of course, were destined for

Facing page
Characteristic TR3A features include a wide grille, Triumph name in separate letters across the nose, and headlamps set further back in the front apron than on previous TRs.

Left
Under the bonnet, little had changed from the TR3. This is the 100bhp (75kW), 2-litre engine in a 1959 TR3A.

export, and of those, the majority were sold in the USA.

Nevertheless, this enormous success was not enough to sustain Standard-Triumph on its own. By 1961, the company was in deep financial trouble. TR3A production was temporarily put on hold, while the company's directors sought help. In due course, they found that help in the shape of Leyland Motors. But TR3A deliveries during 1961 had sunk to around a fifth of their 1960 levels.

In the meantime, the car had undergone a number of specification changes. The most important of these came in 1959, when the rally proved 2.2-litre engine became optional in new cars, while a conversion kit was offered to uprate earlier 2-

Above

On the TR3A, the central facia panel was finished in crackle-black, while the rest of the panel was covered in grained vinyl, as on the TR3.

litre models to the new specification. Although the 2.2-litre engine offered no more power than before, it did develop its maximum power at lower rpm, which improved noise levels and therefore refinement; and it did produce more torque (albeit at higher rpm), which improved acceleration. A TR3A with the 2.2-litre engine was actually 1.5sec faster to 60mph (97km/h) from a standing start than a 2-litre TR3.

At the same time as the larger engine option was introduced, the TR3A's braking system was revised, taking on smaller drums at the rear to improve the front-to-rear balance. Then, in the spring of 1960, a number of detail alterations were made when the bodyshell was retooled. These

Above
In the TR3A, the shapely bucket seats received new pleated upholstery. This view also clearly shows the occasional rear seat option, which in truth was a seat in name only.

Left
The new front-end treatment was the most noticeable difference between the TR3A and its predecessors. However, further minor changes included the addition of exterior door handles.

did not materially alter the appearance of the cars, and mainly affected the area around the cockpit and the windscreen.

For 1962, a thoroughly restyled TR had been developed with the aid of Triumph's regular styling consultant, the Italian Giovanni Michelotti. However, when the North American dealers were shown the car, they expressed serious doubts about whether it would appeal to their customers: it was, they said, too refined. So Triumph agreed to build a limited number of modified TR3As to ease the transition to the TR4 for US customers. The TR3A's body tooling was transferred from the Mulliner plant (which had built all TR bodies up to that point) to the Forward Radiator Company plant in Birmingham, and 3331 TR3A look-alikes, most of them with the all-synchromesh gearbox from the TR4, were turned out in the latter months of 1961. The first 500 of these

TR3B models had 2-litre engines; the remainder had the 2.2-litre type. All were sold in the USA.

There was a third model available during the production run of the TR3A, and in fact this continued to be built until 1963, thus overlapping with the TR4 by two years. This third model was the Triumph Italia, a smart fixed-head coupé styled by Michelotti and built by Vignale on the 2-litre TR3A chassis. Announced in 1958 at the Turin show, it entered series production the following year, wearing Triumph 2000 badges (although, of course, it never had anything in common with the 1963 Triumph 2000 saloon). Since the car was handbuilt, production was limited and prices were high. Most of the 300 or so produced were sold in Continental Europe, and the majority were left-hand-drive models. A few, however, were imported into the UK.

Left
The layout of the engine compartment was pretty much the same as those of previous models. Although apparently original externally, a lack of suitable cylinder liners to original dimensions led the owner of this 1959 TR3A to increase the bore to 87mm, giving a capacity of 2192cc.

Left
In this rear view, note the amber flasher lamps, which had become standard equipment (previously, they had been optional extras). TR3As also had lockable door and boot lid handles, a chromed and underlined 'Triumph' badge on the rear panel, and a chromed number-plate lamp.

Above
The TR3A's hood was still of the 'build-it-yourself' variety. First of all, you erected this tubular framework, which normally lived under the hood bag in a well behind the passenger compartment...

Above right
...then you removed the top itself from the boot and stretched it across the hoodsticks to produce this. It wasn't a quick job and could be a tiny bit frustrating in the pouring rain. Sidescreens, not shown in this picture, completed the ensemble.

Left
This view shows the three-piece rear window of the TR3A's soft top.

Left
This profile of the TR3A clearly shows that the basic lines of the car did not differ from those of earlier TRs; Triumph had simply tidied things up a little.

Above
Another feature that distinguished the TR3A from its forebears was that the standard disc wheels, when fitted, were finished in silver instead of in the body colour. Wires remained optional, of course. Brake drums were normally finished in black rather than the red of this car.

Right
Only TR3As built before January 1959 had red and black enamelled bonnet badges. Normally, the badge came without the Triumph name in the scroll at the bottom, but this is an early car and may well have been fitted with this TR3-type badge from new.

Facing page, bottom
After January 1959, the badge colours were changed to blue and white, seen to good effect on this pale yellow car.

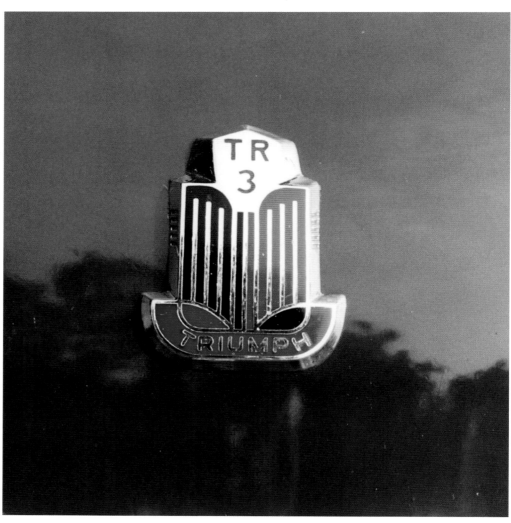

Above
On this left-hand-drive TR3A, the standard black three-spoke steering wheel can be seen, together with the export speedometer with its kilometres-per-hour graduations.

Overleaf, left
In this head-on picture of a TR3A, note the white indicator lamps set into the ends of the grille...

Overleaf, right
...and compare them with the amber lamps fitted to this later example made for the German market. Just as on the home-market car, a wing mirror is fitted to the driver's side only. In both cases, the cars would have had Lucas tripod headlamps when new.

43

TRIUMPH TR3A—SPECIFICATIONS

Engine	Four-cylinder, OHV, carbureted
Capacity	1991cc (83mm bore, 92mm stroke); 2138cc (86mm bore, 92mm stroke) option for 1960 and 1961 models
Max. power	100bhp (75kW) @ 5000rpm; no official figures were released for the 2.2-litre engine, but power is estimated at 100bhp (75kW) @ 4600rpm
Max. torque	117lb/ft (174kg/m) @ 3000rpm; 2.2-litre engine estimated at 127lb/ft (189kg/m) @ 3350rpm
Transmission	Four-speed manual gearbox with optional overdrive
Suspension, front	Independent, with coil springs, wishbones and telescopic dampers
Suspension, rear	Live axle, with semi-elliptic leaf springs and lever-arm dampers
Steering	Cam-and-lever
Brakes	Discs on the front wheels and drums on the rear wheels
Tyres	5.50x15 crossply
Length	12ft 7in (3.84m)
Width	4ft 7.5in (1.41m)
Height	4ft 2in (1.27m) (with soft top erected)
Wheelbase	7ft 4in (2.24m)
Max. speed	102mph (164km/h) (with overdrive)
0-60mph (97km/h)	12.5sec; approx. 11sec with 2.2-litre engine
O'all fuel consumption	25mpg (8km/litre); 2.2 litre engine, approx. 24mpg (8km/litre)
Production total	58,236 (1896 for home market; 56,340 for export)

TRIUMPH TR3B—SPECIFICATIONS

Engine	Four-cylinder, OHV, carbureted
Capacity	Early models, 1991cc (83mm bore, 92mm stroke); late models, 2138cc (86mm bore, 92mm stroke)
Max. power	2-litre models, 100bhp (75kW) @ 5000rpm; 2.2-litre models, 100bhp (75kW) @ 4600rpm
Max. torque	2-litre models, 117lb/ft (174kg/m) @ 3000rpm; 2.2-litre models, 127lb/ft (189kg/m) @ 3350rpm
Transmission	Four-speed, manual all-synchromesh gearbox with optional overdrive
Suspension, front	Independent, with coil springs, wishbones and telescopic dampers
Suspension, rear	Live axle, with semi-elliptic leaf springs and lever-arm dampers
Steering	Cam-and-lever
Brakes	Discs on the front wheels and drums on the rear wheels
Tyres	5.50x15 crossply
Length	12ft 7in (3.84m)
Width	4ft 7.5in (1.41m)
Height	4ft 2in (1.27m) (with soft top erected)
Wheelbase	7ft 4in (2.24m)
Max. speed	102mph (164km/h) (with overdrive)
0-60mph (97km/h)	12.5sec; approx. 11sec with 2.2-litre engine
O'all fuel consumption	25mpg (8km/litre); 2.2 litre engine, approx. 24mpg (8km/litre)
Production total	3331 (all for export)

TR4

The TR4 marked a turning point in TR history. Its restyled body made an obvious break with the tradition established by the TR2s and TR3s, but there were many other reasons why the TR4 was a rather different breed from its predecessors. Undeniably prettier than they were, and undeniably more refined, nevertheless it was not as competitive in the hands of the Triumph works drivers, and did not sell in anything like the same quantities as the TR3A.

The restyled body—longer, wider and heavier than that of the TR3s—was the work of Giovanni Michelotti, the Italian design consultant whom Triumph had placed on a retainer in the late 1950s. It was notable for its full-width styling without the clearly defined wings of the TR2s and TR3s, for its wind-up glass windows instead of detachable sidescreens, for a

much more spacious and well-appointed interior, and for its new style of hardtop.

Triumph called the new hardtop a 'Surrey' top, although the general configuration is more commonly known nowadays as a 'Targa' top. In essence, it consisted of a fixed rear panel, containing a wrap-around rear window, and a detachable roof panel that fitted between the rear panel and the windscreen. The metal roof panel was actually too large to stow in the car, however, so Triumph provided a folding fabric panel as well, which could be carried in the car in case poor weather took its occupants by surprise.

Although the basic chassis configuration of the TR4 was the same as that of the earlier 'sidescreen' cars, there were significant differences in the mechanical specification and running gear. In place of the cam-and-lever steering was a new and

Facing page
Both sidelights and orange indicator lamps were incorporated in the outer ends of the TR4's grille.

Below
The TR4's styling was altogether more modern than the TR3A's. The white hood fitted to this car was an optional factory fitment and makes a striking contrast with the Signal Red paint, although it was not easy to keep looking as clean as this one does! This car is a left-hand-drive example, which was imported into the UK.

Below

The Michelotti styled TR4 was very different from earlier TRs, most noticeably because of its slab sides in place of the clearly defined front and rear wings of the TR2s, TR3s and TR3As. Yet the wing-line did not run straight through, for there was a little kick-up at the rear of the door to give the side elevation some character. The nose had also been completely restyled, but Michelotti had managed to preserve the flavour of the earlier TRs. The bonnet bulge was not just for show: it was necessary to clear the twin SU carburettors, which still sat high up, even though the body had been lowered around them.

much more positive rack-and-pinion type, and in place of the crash-first gearbox was a transmission with synchromesh on all forward gears; overdrive, of course, remained optional. The 100bhp (75kW) 2.2-litre engine had become standard, but the extra weight of the new body prevented the TR4 from being any faster than its predecessors, and also made its fuel consumption slightly worse.

The TR4 was in production for just over three years. During that period, it changed very little. New front suspension geometry and revised brakes arrived early in 1962, while the summer of 1963 saw the introduction of new seats and Stromberg carburettors instead of the SUs fitted previously. Both of these latter changes had been seen on a trial batch of 100 cars built during the winter of 1962-63. But work had been under way on an even more sophisticated TR since 1962, and the TR4 gave way to this, the TR4A, early in 1965.

Between 1962 and 1964, Triumph ran a works team of four light-alloy-panelled TR4 coupés. However, they were not strong enough to be competitive in the increasingly tough long-distance rallies of the time, and not fast enough to be successful against increasingly rapid competitors on tarmac. These TR4s were the last traditional TRs to appear in the works competition team: during 1964, they were replaced by Triumph 2000 saloons on long-distance rallies, and by Spitfires for Le Mans and other tarmac events.

One variant of the TR4 deserves special mention here. This is the Dove GTR4, a fastback coupé version, which was built between 1961 and 1964 by Harrington's for the Triumph dealer L. F. Dove of Wimbledon. The styling of the coupé top was not to everyone's taste, however, and the additional weight of the car (some 500lb/227kg of it) took the edge off performance. Only 55 GTR4s were built.

Left
Michelotti gave the TR4 a much larger boot than its predecessors. It was more conventional, too, reaching down to bumper height and opening to reveal a flat floor, under which the spare wheel was stowed. TR4s were the first TRs to have separate lettering for the Triumph name on the boot lid, and were also the first Triumph-built TRs to have a proper rear bumper.

Below
Earlier TRs had sidescreens, but with the TR4 came wind-up windows. There was no longer any need for the door tops and rear of the cockpit to be upholstered.

THE TESTERS' VIEW

Facing page, top
Early TR4s used the same seats as the TR3A, as this picture shows.

Facing page, bottom
The optional rear seat was only suitable for very occasional use, as it had been since the days of the TR2, but it was neatly padded.

Below
With the TR4 came the first wooden dashboard for a TR, but only for the US market, and only at the end of TR4 production. The wood-rimmed steering wheel fitted to this car is not an original feature; TR4 steering wheels were finished in black plastic.

So marked are the differences in appearance between the new Triumph TR4 and the old series, now discontinued after production through seven years with hardly any major changes, that it is inevitable to think of the latest version as being entirely new. In fact, it still has basically the same chassis, and TR enthusiasts will find that in its fresh guise the Triumph sports car retains most of the familiar characteristics of earlier models, supported by many useful improvements.

Without doubt the greatest advance shown by the new car in comparison with the TR3A is in the more shapely and practical design of the body, and it is commendable that in comparison with the TR3A hardtop, which we tested just five years ago, the new model shows a weight saving of over 2cwt *(102kg)*. With this goes a 7½ per cent increase in engine capacity—to 2,138 c.c., so that the new car is quicker off the mark than its predecessor. In acceleration from rest, 60 m.p.h. *(97km/h)* was reached in 10.9sec, compared with 12.5sec for the TR3, and acceleration to 90 m.p.h. *(145km/h)* was improved from 33.8 to 28.2sec. The time for the standing quarter-mile *(0.4km)* is now almost a full second quicker, at 17.8sec.

In timing this test, incidentally, the consistent performance and evenness of the torque of the TR4's lusty four-cylinder engine was emphasized by the fact that it recorded identical standing quarter-mile times to the nearest tenth of a second, four times in succession in opposite directions. The extra engine capacity is coupled with an increase of compression ratio to 9 to 1, and in this form the engine develops 5 b.h.p. *(4kW)* more than the earlier version, and peaks at lower revs—4,600 instead of 5,000 r.p.m. The extra torque is the more noticeable advantage, and in particular there is much more eagerness from the engine below 2,000 r.p.m., thus extending the usable range of engine performance.

Laycock-de Normanville overdrive is still available but was not fitted to the test car.

With the 3.7 axle used, the car runs at 20 m.p.h. *(32km/h)* per 1,000 r.p.m. in top gear, and hence at 100 m.p.h. *(161km/h)* the engine is well beyond the peak of the power curve. In fact, the TR4 will not exceed 5,000 r.p.m.—given as the upper limit for engine safety—in top gear on level ground, and the best one-way speed of 104 *(167km/h)* was the maximum at which the car "ran out of steam"; the 4 m.p.h. *(6km/h)* increment over the 100 was accounted for by tyre growth. Cruising at 90 m.p.h. *(145km/h)* is permissible without exceeding the recommended engine rev limit of 4,500 r.p.m. for sustained use.

Exhaust noise outside the car is not obtrusive, but a fair amount of roar is heard from within, and this combines with the high level of wind noise to make the car almost as noisy to travel in as its predecessor. Twin S.U. carburettors are fitted, and there is an induction hiss when the throttle is opened. When starting from cold, little or no choke is needed, and apart from an initial tendency for the engine to "die" if the throttle is opened abruptly at low revs, the warm-up is quick. The test car evidently was equipped with a relatively "cold"

Right
Like the TR3A, the TR4 had its standard disc wheels finished in aluminium colour, regardless of the colour of the body. The familiar Triumph 'globe' still featured on the hub-caps.

Below
This TR4 in Conifer Green shows the fixed rear window of the optional 'Surrey' top.

thermostat, even when allowance is made for the fact that the temperature gauge take-off is not in the cylinder head, and the instrument seldom indicated more than 150 deg F *(66°C)*. With better temperature maintenance a small improvement in fuel economy might be obtained.

Throughout the test, the best consumption figure returned was 26.3 m.p.g. *(9km/litre)* on a main road run where full acceleration was used on occasions for overtaking, but with many miles at a low speed enforced by dense traffic. The TR4 obviously is a car which will prove economical for those who are prepared to drive quietly, but for the majority of owners who will want to take advantage of the eager acceleration and high-speed cruising, 22 or 23 m.p.g. *(8km/litre)* will be a more normal consumption figure.

The capacity of the fuel tank has been reduced from the 12.5 gallons *(57 litres)* of the TR3A to 11.75 *(53 litres)* with the TR4, so that with the heavier consumption the previously long range between refuelling stops has been lost. A large fuel filler with snap-up cap allows petrol to be delivered at the maximum rate of a modern pump since it opens directly to the tank, and it is even possible to make a visual check of how much remains should the contents be very low. The fuel gauge is adequately damped to give a steady reading, but on the test car it was inaccurate, reading "half" when the tank required only four gallons *(18 litres)* to fill.

No difficulty is experienced in making a clean start from rest with use of full power, as the clutch pedal can be released swiftly with the engine turning at high revs without provoking wheelspin or any unwanted clutch slip. A hydraulic clutch linkage is used as before, but the pedal travel is excessive and is noticeably heavy when driving in city traffic.

Well-chosen gear ratios have always been a good point of the TR series, and are retained in the TR4 with the commendable advantage of synchromesh on all four gears. This allows bottom to be engaged right up to 30 m.p.h. *(48km/h)* without any attempt at double-declutching, but the chief advantage is that when the car is stationary this gear may be engaged silently at once. On the test car the change was stiff to move, but it was felt that this would free with further use, as the car was received with less than 2,000 miles *(3219km)* behind it. The change is neverthe-

less delightfully precise to use; the lever itself is cranked back, bringing it within easy reach of the driver. Considerable gear whine is audible when the indirect gears are in use, but this, too, may diminish with greater mileage.

Normal change-up speeds are about 20, 40 and 60 m.p.h. *(32, 64 and 97km/h)*, but if the full performance is being used, advantage can be taken of the usefully high maxima of 33, 51 and 77 m.p.h. *(53, 82 and 124km/h)* in the three indirect ratios. An alternative final drive ratio of 4.1 to 1 may be specified, though there seems little call for this lower effective gearing except possibly in conjunction with the overdrive. What would be appreciated would be an axle offering higher effective gearing, to give some degree of the benefits of overdrive without the associated initial expense.

Rack-and-pinion steering of similar pattern to that used on the Triumph Herald is featured on the TR4, and has the same safety provision for the column to telescope in a frontal accident, avoiding steering wheel injuries to the driver. The steering itself is exceedingly good, and although TR steering has always been satisfactory the new rack-and-pinion unit shows gains in both accuracy and lightness. On smooth road surfaces the car's direction at high speed is controlled by delightfully small steering wheel movements.

Unfortunately no appreciable advance has been made where the suspension is concerned,

Above
A new type of chassis number plate was used on TR4s, and was fitted on the opposite side of the car from that fitted to earlier TRs. The commission number of this car—CT 9211 LO—reveals that it is a 1962 model, built with left-hand-drive (L) and the optional overdrive (O).

Right
The bonnet badge of the TR4 was the same shape as that of earlier cars, and was still finished in blue and white enamel. But there was no Triumph name on the lower panel, because the marque name was spelled out across the nose in separate letters.

Right
This close-up shows the large chromed 'TR4' identifier on the boot lid. That 'pip' on the inner surface of the overrider is actually one of the number-plate lamps.

and the ride is still decidedly harsh, with the result that the wheels tend to hop over any but the smoothest of surfaces. When cornering this results in a degree of rear-end steering which requires constant correction, and the fault is accentuated if the harder tyre pressures recommended for fast driving are used. On really bad roads, such as Continental pavé, severe bottoming of the rear suspension occurs, and the car makes violent vertical movements.

There may be many who would feel that a "vintage" suspension of this kind is appropriate to the sporting character of the TR4, while in any case the ride is satisfactory on well-maintained roads and tends to flatten out at high speeds. Fairly marked understeer contributes to good directional stability, though on severe undulations taken at speed the TR4 tends to float and settle slightly askew. Cross winds have considerable effect on the car's

direction, but a reasonably straight course is easily held.

In hard cornering the rear wheels break away fairly readily, balancing to some extent the understeer tendency, and provided cornering speeds are kept to reasonable levels there is no reason for any driver to miscalculate the cornering abilities of the TR4. The extra 4in. *(10cm)* on the track of both front and rear wheels has contributed towards making the handling characteristics more predictable than they were with the earlier model, but the car still slides readily in the wet.

Dependable brakes

It may be recalled that the 1957 TR3—prior to the TR3A— was one of the first production cars to feature disc brakes on the front wheels, and these Girling brakes are unchanged for the TR4. Heavy pedal pressures are required for maximum applications, but the brakes are notably progressive and thoroughly dependable. When used hard, they have real "bite", particularly above about 60 m.p.h. *(97km/h)*, and on test they proved capable of a reassuring 0.95g stop. On wet roads they can be used quite hard without fear of locking the wheels, while no amount of water pick-up in heavy rain appears to have any effect on their efficiency. The handbrake is controlled by a fly-off lever to the right of the transmission hump; it held the car securely on a 1-in-3 gradient.

Beside the handbrake lever is a substantial reinforcement bracing the gearbox cover and floor structure direct to the facia, and this has contributed to a commendable degree of rigidity as far as the scuttle is concerned, but some squeaky flexing of the bodywork and movement of the bonnet are noticed on rough roads. The use of frameless winding side windows has helped to keep the obstruction of the windscreen side pillars to an acceptable minimum, in keeping with the generally high standard of visibility. The steering wheel and scuttle are pleasantly low, and both of the arched front wing peaks are visible from the driving seat. The headlamp hoods formed at the lip of the bonnet, and the "power bulge" enabling a low bonnet to clear the carburettors, are also seen, combining to make rather a confusion of humps within the driver's view.

Self-parking single-speed wipers clear the screen well in front of the driver, and do not lift at high speed. The left wiper pivot could well be moved outwards to reduce the large unswept portion in front of the passenger, as at present the gap between the sweep of the blade and the left windscreen pillar is seven inches *(18cm)* at the nearest point.

The driving position generally is good, and is aided by the fact that the steering wheel may be repositioned fore and aft to suit a particular driver; the pedals are well-placed and allow easy combined operation of brake and accelerator with the right foot. A marked disappointment for owners of the earlier models is that the support provided for the left foot is considerably nearer to the driver, who must now sit with the right leg extended to the throttle and with the left leg bent except when operating the clutch. On a long journey this can become decidedly uncomfortable, even tempting the driver to insinuate his left foot underneath the clutch pedal to enable him to stretch the leg. This forward positioning of the

rest may have been adopted to avoid offsetting the pedals to the right, and it does mean that the left foot slides easily from the rest on to the clutch at the same level.

The seats are adjustable fore and aft over a wide range, and they give good lateral support provided their occupants are reasonably slim enough to be accommodated by the marked curvature of the backrest; and the cushions extend well under the thighs.

Provision is made for the passenger seat backrest to fold forward, for reaching the subsidiary luggage platform behind the seats and the hood stowage. The new hood mechanism is now a vast improvement over that of the earlier model, and when in position it provides complete weather protection even in the storm conditions once experienced during the road test. Raising and lowering the hood, however, remains quite a procedure, and when the car is

Right
A rare survivor: the Harrington-bodied Dove GTR4. Only 55 were made during the period 1961-64.

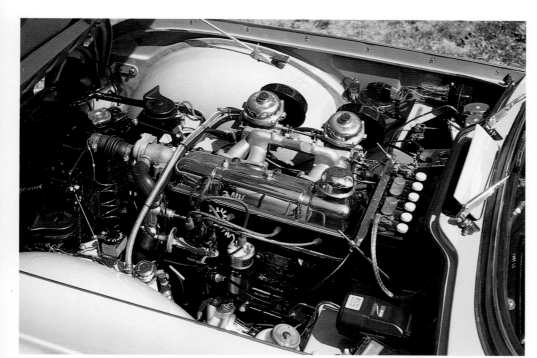

Left
Under the TR4's bonnet was still the 100bhp (75kW) 2.2-litre engine, which meant that the new car was not noticeably faster than the TR3A it superseded.

Right
Harrington Motor Bodies (better-known for their coach bodywork removed the boot and rear deck of the TR4, then replaced them with a full-length moulded GRP roof section incorporating a hinged tailgate.

Below
The GTR4, like most of the standard TR4s, had a metal dashboard, painted white, with a black crackle finish centre panel. The general layout was similar to that in the sidescreen cars, but with a large speedometer and rev-counter directly ahead of the driver, and four more gauges in the centre of the dash.

open the top itself is an awkward commodity for stowing in the boot with any luggage being carried. Packing it beneath the rear flap with the support rails is not recommended. An ingenious over-centre lever in each main support allows the framework to be tautened after the material is fastened.

A full-length tonneau cover is available for when the car is parked with its hood removed, and either open or closed the TR4 certainly looks neat and functional. In this respect it is such an advance over its predecessor that it now makes this and some of its competitors with detachable sidescreens look decidedly out of date. A similar advance is shown by the heating and ventilation arrangements. The heater is extra, priced at £16 10s (£16.50) including tax, and as fitted to the test car it provided ample delivery of warmed fresh air drawn from a vent on the top of the scuttle. A single-speed fan is provided to boost the air flow at low speeds.

Below

Like the optional occasional rear seat in the open TR4s, that of the GTR4 was very restricted for passenger legroom. However, it could be folded forwards to provide extra baggage space at the rear. The four-point competition seat harnesses were not standard equipment.

For use in warm weather, an adjustable cool air vent is provided at each end of the facia. These vents are an admirable provision, and may be used to blow cool air on to the face when the heater is in use in winter, but there appears to be no trap to prevent insects or grit from blowing through the tunnels. Instrumentation is first class, with the rev counter and speedometer (with trip mileometer) set in large dials in front of the driver and matched on the left by a lockable glove compartment. The minor instruments are grouped neatly on a bright alloy background in the centre of the facia. Variable control is provided for the illumination, but unwanted light from the instrument panel edges irritated some drivers, even with the intensity fully dimmed.

Just below the steering wheel rim is a convenient finger-tip control for the winking indicators, which flash amber at both front and rear. This control is matched by the overdrive switch when this accessory is fitted, but there

is no provision for flashing the headlamps. Lucas 700 headlamps are fitted, and as they are mounted fairly low they have a particularly short throw when dipped; they also lack the intensity of main beam necessary for driving at high speeds. The foot dipswitch is awkwardly placed. There is no reversing lamp provision, nor any interior light.

Accessibility afforded by the new forward-hinged bonnet is excellent, and will be welcomed especially by the many enthusiast owners who will want to apply their own attentions to the power unit, but it is a pity that no attempt has been made to reduce the need for 16 chassis points to be greased every 1,000 miles *(1609km)*. The bonnet is held up by a self-fixing stay, though a similar prop is not provided for the boot lid, which requires two free hands to secure it.

Despite some shortcomings, the new Triumph TR4 is an invigorating car to drive, offering eager performance with compactness and manoeuvrability. It is likely to appeal to much the same section of the sports car market that was attracted by its predecessors, and those enthusiasts will no doubt be joined by others to whom the greatly improved appearance and all-weather equipment will appeal.

Reprinted from Autocar, 5 January 1962.

Below
Folding the GTR4's back seat down provides a useful flat load carrying platform. The electrical contacts visible at the bottom of the picture are for a non-standard rear window wash/wipe system added in more recent years.

Left
The GTR4 was considerably more expensive than the standard TR4, and no doubt this was responsible for its lack of success. However, it did point the way for more practical sporting coupés in the future. This example has been fitted with a rear window washer and wiper, and fog and reversing lights to cope with the demands of modern motoring.

TRIUMPH TR4—SPECIFICATIONS

Engine	Four-cylinder, OHV, carbureted
Capacity	2138cc (86mm bore, 92mm stroke); 2-litre TR3A engine optionally available on early cars
Max. power	100bhp (75kW) @ 4600rpm
Max. torque	127lb/ft (189kg/m) @ 3350rpm
Transmission	Four-speed, manual all-synchromesh gearbox with optional overdrive
Suspension, front	Independent, with coil springs, wishbones and telescopic dampers
Suspension, rear	Live axle, with semi-elliptic leaf springs and lever-arm dampers
Steering	Rack-and-pinion
Brakes	Discs on the front wheels and drums on the rear wheels
Tyres	5.90x15 crossply
Length	12ft 9.6in (3.91m)
Width	4ft 9.5in (1.46m)
Height	4ft 2in (1.27m) (with soft top erected)
Wheelbase	7ft 4in (2.24m)
Max. speed	102mph (164km/h)
0-60mph (97km/h)	10.9sec
O'all fuel consumption	22.5mpg (8km/litre)
Production total	40,253 (2592 for home market; 37,661 for export)

TR4A

In the TR4, Triumph had radically updated the appearance and the creature comforts of the TR range. But almost as soon as that model had gone into production, the company had started work on the next stage of the TR transformation, which was to take the form of an updated chassis.

Independent rear suspension was by no means new to Triumph, since the company had introduced it as early as 1959 on the small Herald saloon, and had made it an essential part of the larger 2000 saloon's specification. However, neither of the TR's natural rivals, the MGB and Sunbeam Alpine, enjoyed such a feature, so by developing an independent-rear-suspension TR, Triumph was assured of a

valuable technological lead which, it was hoped, would sell more cars.

For the TR4A, introduced at the beginning of 1965, the entire TR chassis was completely redesigned. It was made more rigid than before and incorporated a semi-trailing-arm independent rear suspension, with coil springs and lever-arm dampers. As a result, roadholding and handling were both vastly improved over the earlier cars, although there was a manufacturing cost penalty.

In the USA, Triumph dealers were unwilling to pass this on to their customers, so Triumph built them a special US-only TR4A with a traditional leaf-sprung live rear axle mounted to the redesigned chassis. The all-independent

Facing page
Triumph had squeezed a little more power from the 2.2-litre engine by fitting a new camshaft, but the under-bonnet appearance of a TR4A was much the same as it had been in earlier cars.

Below
From behind, the TR4A looked just like a TR4, unless you could see the badging. The new independent rear suspension and wider tracks were not at all obvious. This car is fitted with the optional knock-off wire wheels.

Right
Although the overall appearance of the TR4A was the same as the TR4, the sidelamps were no longer incorporated in the grille and there was a round medallion above the Triumph name instead of the shield type.

TR4A was also available in the USA, albeit at extra cost.

Other changes from the TR4's mechanical specification were few. A new camshaft and a twin-pipe exhaust system helped to extract a little more power from the 2.2-litre engine, but the TR4A had again put on weight and could not offer any performance gains over its predecessor. Minor improvements included a diaphragm-spring clutch and the option of an alternator in place of the traditional dynamo.

The TR4's Michelotti styled body still looked modern, so Triumph retained it almost unchanged for the TR4A. To distinguish one model from the other, they gave the TR4A a new grille, new badging front and rear (including a proud and prominent 'IRS' at the back), and chrome flashes leading back from new indicator repeater lamps on the front wings. In the passenger compartment, changes were also minimal. TR4As took on the later type of TR4 seat (albeit trimmed in a different type of vinyl), but the dashboard was more sumptuously finished in walnut veneer, and the handbrake was made more accessible by locating it on top of the transmission tunnel. Unlike their predecessors, TR4As were not available with optional occasional rear seats, and the space behind the front seats was always trimmed as a luggage area.

Despite all these improvements, the TR4A did not prove as popular as the TR4, which itself was less popular than the TR3A. It was quite clear to Triumph that a more radical redesign would be necessary for the next generation of TRs if the range was to keep its competitive sales edge.

Facing page, bottom
The 'Surrey' top option fitted to this TR4A was first introduced for the TR4. In place of a conventional hardtop, it offered a fixed rear window unit and detachable roof panels. A metal section gave the appearance of a proper hardtop...

Below
...while a folding fabric section made the car feel more like a soft-top. Note the panoramic rear window, which offered exceptionally good rearward vision.

THE TESTERS' VIEW

If a car is built to a basically sound formula and the designers apply improvements to it by a process of progressive development, then the demand for it will continue long after new, more elegant rivals have come on the scene. The Triumph TR series has always provided a car in the traditional British image for people who enjoy motoring and prefer not to be completely insulated from the car—or even the elements at times. Now that the emphasis generally is switching from performance to more creature comfort, a new model has replaced the TR4 to bring its specification into line.

A lot of the TR formula remains unchanged—separate chassis, lusty 4-cylinder engine, bolt-on wings, and even a detachable windscreen—but a new fully independent rear suspension has been added, together with a more refined interior having the latest trimming materials and a touch of woodwork. Yet the recommended total price is only up by £60 on the previous model, and it is still possible (just) to drive the latest version away from the showroom for an outlay of under £1,000.

For some time now Triumph have been using Stromberg CD constant-vacuum carburettors instead of twin S.U.s, and with the introduction of the TR4A in March this year they altered the camshaft timing slightly and fitted a new, cast-iron exhaust manifold and dual exhaust system to reduce the noise level to the new international standards. At the time it was claimed that these changes also gave the engine 4 b.h.p. *(3kW)* more at its peak of 4,700 r.p.m. and increased maximum torque from 127 to 132 lb. ft *(189 to 196kg/m)* at 3,000 r.p.m. In fact, the performance of the latest car is only marginally different from the previous one tested, as most of the extra power is "lost" in overcoming the additional weight of the improvements and overdrive.

Performance

As tested before, the TR was without overdrive, so for direct comparison the extra three ratios on this car should be ignored. Top gear speed therefore is identical—103 m.p.h. *(166km/h)* mean, 104 *(167km/h)* best. From a standing start (using overdrive ratios where necessary to get the best times) the TR4A is 0.5sec slower to 60 m.p.h. *(97km/h)* (11.4 compared with 10.9sec) and 2.5sec slower to

100 m.p.h. *(161km/h)* (48.8 compared with 46.3sec). However, measured over 20 m.p.h. *(32km/h)* increments in top gear the 4A is quicker by the odd half second or so in the middle of the rev range, but loses out from low speeds and at the top end.

Judged on its own, especially across country, the TR4A is remarkably brisk. Overdrive helps considerably in gaining speed, for it works on all gears except first. Direct top has practically the same ratio as overdrive third, giving in effect a close-ratio 6-speed gearbox; the o.d. selector switch is arranged conventionally within fingertip reach of the right hand so that it can be used simultaneously with the gearlever. Working up through this "double" gearbox the maximum speeds are 33, 51, **60**, 76, **98** and **109** m.p.h. *(53, 82, 97, 122, 158 and 175km/h)* (o.d. ratios in bold). It would be better if the o.d. switch were self-cancelling or at least incorporated a warning lamp, for one can sometimes get confused

with the arrangement and change from a direct gear to the next higher o.d. (particularly from first to o.d. second) with a substantial loss of revs and hence a lag in acceleration.

The overdrive itself is one of the new designs announced last autumn and works with quite outstanding smoothness. A little throttle is best for both upward and downward changes, and then the device engages quickly without any jerkiness—more smoothly, in fact, than most drivers can manipulate a clutch pedal. Not only does the higher top gear ratio increase the maximum speed of the car by some 6 m.p.h. *(10km/h)* but it brings 100 m.p.h. *(161km/h)* down to only 4,000 r.p.m.— a speed the car can hold all day if required without strain, stress or fuss. Fuel consumption is improved by almost 20 per cent with the aid of overdrive, and at 100 m.p.h. the TR managed 19.2 m.p.g. *(7km/litre)* in o.d. top (compared with 16.2 m.p.g. *[6km/litre]* in direct). Overall we recorded 25.4 m.p.g. *(9km/litre)*, which is exceptionally good for a 2.2-litre sports car.

The characteristics of this Triumph engine are well known; it is a big four with three main bearings and a rather long stroke by today's standards (86x92mm). Maximum revs are about 5,000 although with the car open one can easily run to 5,500 without noticing. At about 3,000 r.p.m. there is a slight rumbling vibration and the engine feels harsh when pulling hard. There is a lot of torque and, if one wants to, the car will pull away from as low as 1,250 r.p.m. (25 m.p.h. *[40km/h]*) in top without snatch or pinking on

super premium fuel. This engine runs quite well on ordinary premium, but there was some running-on once or twice, so we settled for a half-and-half mixture of the two top grades.

With the new exhaust system there is no question of the car offending neighbours in the small hours, and it is a good deal quieter from the outside than one would expect from the driving seat. Inside, one hears a lot of induction roar above the other mechanical noises, and with the hood up at speed there is considerable wind howl. There is no flapping from the fabric, which always remained taut and, as far as we could judge in a couple of isolated thunderstorms, watertight too.

The 4-speed all-synchromesh gearbox is unchanged. On our car, which had completed only 3,000 miles, it was somewhat stiff with a notchy action, but we know from other cars that this wears off after use. A diaphragm-spring clutch is now fitted—it has the same loading as previously but an inch *(2.5cm)* less movement. We found its efficiency only marginal for a car of this performance, excessive slip being all too easy to provoke during fast getaways and a 1-in-3 being too steep for it to grip at all when we tried a hill restart.

Tyres

Cars destined for export can be ordered with French-made Dunlop SP Sport tyres, which look like the now-unobtainable British SP3s, although for the home market Goodyear Grand Prix are standard with the option of Dunlop SP 41 or Michelin X. The test car was shod with these French tyres, which gave

Right
The combined sidelamp/indicator units were perhaps rather fussy for the mid-1960s, but they did help to make the TR4A distinctive.

Left

Another minor difference between the TR4A and TR4 was the addition of trim to the door tops of the later cars. On TR4s, the inner surfaces of the door tops were simply painted metal.

Left

The TR4A was the first of the TRs to dispense with the traditional enamelled shield badge, and wore instead a version of the Triumph 'globe', which was still in evidence on the hub-caps of disc-wheeled TR4As. The disc wheels themselves came with an aluminium finish, as on previous TRs.

Left

The TR4A's rear badging proudly proclaimed 'IRS'— Independent Rear Suspension. And it was something to be proud of, for the only other mass-produced sports cars with independent rear ends when the TR4A was introduced were Triumph's own Spitfire and the much more expensive E-type Jaguar The reversing light fitted to this car is a period accessory, but was not standard.

superb adhesion under all conditions. Largely because of this advanced tyre we were able to send the Mintex manometer right off its scale when measuring brake efficiency at 30 m.p.h. *(48km/h)*. This was at a pedal load of only 100lb *(45kg)*, the car pulling up fair and square with no squeal or wheel-lock. So good was the grip that it took an extra 50lb *(23kg)* load on the pedal before all four wheels locked together, again without tyre squeal.

Still with the traditional fly-off action, the handbrake has been moved from the side of the transmission tunnel to on top, between the seats. Unfortunately, there is less leverage now and although we were able to record 0.4g from 30 m.p.h. the car was not secure on anything steeper than 1-in-4.

Ten stops from 70 m.p.h. *(113km/h)* showed a slight degree of fade, but the car still stopped at 0.5g with only 65lb *(29kg)* load on the pedal and there was no sponginess or feeling of reduced efficiency.

The greatest improvement in the TR comes from the new rear suspension. This is similar to that used on the Triumph 2000 with coil springs and cast light-alloy semi-trailing arms. Because there is much less unsprung weight to control and the springs no longer locate the axle, the suspension is much softer and better able to cope with rough roads. Longer front springs give more suspension travel, and consequently the ride characteristics have changed out of all recognition. Before, one could never forget the live back axle, which pattered and thumped about, especially if disturbed in a corner taken under power. Now the TR can be driven deliberately fast at obstacles it would have shied from before.

On motorways the undulating waves from subsidencies cause no recurrent pitching and it was only on our closely spaced undulations at M.I.R.A. that we were able to get the back to kick up in the air. On pavé there is a good deal of bottoming, but the car is not thrown from side to side and keeps a very true path.

Road Holding

Cornering is transformed, and there is now a steady degree of understeer, with quite a lot of body roll. If power is applied early in a bend,

Right
From the side, the TR4A is almost indistinguishable from the TR4—only the combined front sidelight/indicator units give it away.

eventually the tail will slide out, but this does-n't happen until well beyond normal speeds. Sometimes, when taking the double bends of a roundabout for instance, there seems to be a slight rear wheel steering effect which upsets one's chosen line and causes the car to go through with a kind of rolling choppiness.

Part of this lack of precise cornering is due to the steering, which now has $3\frac{1}{2}$ turns from lock to lock instead of $2\frac{1}{2}$. It is considerably lighter than before, and it calls for no great effort to wind between locks with the car stationary, as when parked in a very small space. However, at speed there is a certain delay in response to slight movements which seems to stem from the rubber mounting of the steering rack. Side gusts tend to displace the car noticeably.

At maximum speed in a straight line the car is much more stable than the TR4 and all the previous wandering has been eliminated. Once the car is set up on a steady radius, such as a fast motorway curve, it holds its line accurately.

The new seats are upholstered in stretch

Above
The walnut dashboard gave the car a more sumptuous feel than previous TRs, but otherwise the interior was basically the same as the TR4. The area behind the seats could no longer be specified with an occasional bench seat.

Facing page, top
The overall appearance of the TR4A belied the changes that had been made beneath the skin, the most significant being the adoption of independent rear suspension. Although well received by the press, its sales were disappointing.

Facing page, bottom
The TR4A was a comfortable touring car, and with the addition of a boot lid luggage rack could carry a useful amount of baggage. This left-hand-drive model, belonging to a German owner, has been fitted with an aftermarket steering wheel.

p.v.c. and are mounted quite high so that the driver's legs are well bent unless he has the clutch on the floor; our taller testers felt rather perched up. Somehow, these seats do not feel completely compatible with the controls, and most of us would have liked more support under the thighs and a longer, more steeply raked backrest (the angle is fixed). As it is, one sits too close to the large (16in. *[41cm]* dia.) wheel and feels somewhat cramped, with the window sill rubbing the right shoulder.

Most of the switches are in the same place as before, except for the lamp control which is now a stalk to the left of the wheel. This moves through three quadrant positions for off, side and headlamps, and pulling it towards the wheel rim flashes the main beams. The foot dipswitch is still retained, however, which is a pity as it is difficult to find and not easy to use; it would be better if the other Triumph system of a remote master switch were employed, with the stalk acting as a selector and dipswitch.

The new walnut veneer dashboard is very well fitted and stands up admirably to Californian sunshine, we are told. It is slightly marred by two clumsily cut holes for warning lamps between the main instruments, but otherwise gives a nice touch of luxury.

Our car was the convertible with a folding p.v.c. roof, as opposed to the more expensive fixed head coupé which has a detachable central roof panel. The design of the hood is one of the best we have come across in any class of car; it is only a matter of moments to lower or raise it single-handed. When stowed a neat cover keeps everything out of sight.

At night the brilliance of the instrument lighting can be controlled by a rheostat with just the right range. The headlamps severely restrict one's speed on dipped beam. There is no reversing lamp, but the twin number-plate lamps light up the inside of the boot very well with the lid open.

The lasting impression of the TR4A is that of a very comfortable touring car which is tough and rugged and well able to storm the roughest roads. It is certainly the sort of car for driving to the Mediterranean in, and having fun with when you get there. The engine is a willing beast, with the same qualities as an eager young farm horse rather than those of a racing thoroughbred—and something of the same ruggedness about it too; yet it pulls the car easily to over 100 m.p.h. *(161km/h)* with remarkable economy and promise of long life.

Reprinted from Autocar, 28 May 1965.

TRIUMPH TR4A—SPECIFICATIONS

Engine	Four-cylinder, OHV, carbureted
Capacity	2138cc (86mm bore, 92mm stroke)
Max. power	104bhp (78kW) @ 4700rpm
Max. torque	132lb/ft (196kg/m) @ 3000rpm
Transmission	Four-speed, manual all-synchromesh gearbox with optional overdrive
Suspension, front	Independent, with coil springs, wishbones and telescopic dampers
Suspension, rear	Independent, with coil springs, semi-trailing arms and telescopic dampers. For USA only: optional live axle, with semi-elliptic leaf springs and lever-arm dampers
Steering	Rack-and-pinion
Brakes	Discs on the front wheels, drums on the rear wheels
Tyres	6.95x15 crossply
Length	12ft 9.6in (3.91m)
Width	4ft 9.5in (1.46m)
Height	4ft 2in (1.27m) (with soft top erected)
Wheelbase	7ft 4in (2.24m)
Max. speed	109mph (175km/h) (with overdrive)
0-60mph (97km/h)	11.4sec
O'all fuel consumption	25.4mpg (9km/litre)
Production total	28,465 (3075 for home market; 25,390 for export)

TR5 & TR250

With the TR4, Triumph had updated the body of the TR sports car; with the TR4A, the company had improved the chassis; now, with the TR5, at long last the performance of the TR range had been uprated by substituting a six-cylinder engine for the old wet-liner four-cylinder type.

The new six-cylinder engine was developed from the l998cc type first seen in the Standard Vanguard of the late 1950s, and then in revised form in the Triumph 2000 saloon. As there was no room to enlarge the bore size, Triumph increased the stroke, ending up with a swept volume of 2.5 litres. By combining a fairly wild camshaft with Lucas fuel injection, the development engineers found that they could produce low-speed driveability with high performance, and in this form the new engine was dropped into a mildly modified TR4A chassis to produce the new TR5.

Unfortunately, the Lucas fuel injection system could not be made to meet the new North American exhaust emissions regulations, which were due to affect 1968-model cars. As a result, Triumph had to develop a separate, emissions-controlled version of the 2.5-litre engine for the US market. This had twin Stromberg carburettors and developed 104bhp (78kW)—the same as the four-cylinder TR4A—as against the 150bhp (112kW) of the fuel injected engine. As a result, the six-cylinder cars for the US market were barely

Facing page
The boot lid of the TR5 was positively busy with badges. The lower plate proclaimed that the car had the fuel-injected engine (Triumph always called it 'petrol injection', hence the 'PI').

Below
Although the TR5 looked very similar to the TR4A, the rear wing badging and repeater flasher lamp were instant recognition features. None the less a distinguishing feature was the lock barrel below the door handle.

quicker than their predecessors, and the main reason for changing to the larger engine was completely nullified. As the carbureted US car was so different in character to the injected model, Triumph decided to give it a special model name, thus US buyers were offered a TR250 while the rest of the world could buy a plain TR5.

To ensure that the 120mph (193km/h) injected cars had adequate roadholding and stopping power, Triumph gave them wider wheels and larger front brake discs than the TR4A, and standardized a brake servo. Both carbureted TR250 and inject-ed TR5 models had these features, but only the TR5 was given a stronger rear axle with raised gearing; the TR250 had the old TR4A component. Fortunately for Triumph, their US dealers agreed to take

all TR250s with independent rear suspen-sion, so that the company did not have to build an alternative live-axled car, as it had done with the TR4A.

The new US regulations for 1968 also covered safety related items, and the changes that Triumph made to meet them were standardized on both TR250 and TR5 models. Most of these changes affected the interiors of the cars, where window winders, door handles, switches and steering wheels were padded or made softer, and the dashboards were given padded edges.

Of course, all these changes caused increases in manufacturing costs, and some of these had to be passed on to the customer. As the cars were little different cosmetically from the TR4As, Triumph adopted radical measures to ensure that

Right
With extra badging and repeater lamps, Michelotti's design was beginning to look a little cluttered, and it would not be long before a restyled TR would appear. Consequently, only 1161 TR5s were built for the UK market, making it one of the rarest of the early TRs.

sales remained healthy. On the TR250, they applied garish 'speed stripes' around the nose of the car, and on both the TR250 and TR5, a heater was made standard for the first time. Also garish were the Rostyle wheel covers, which were standard equipment on both cars, although fortunately many owners preferred the wire wheel option.

Soon after the TR250 and TR5 had gone into production, Triumph's management decided that a restyle would help sales. As a result, the cars were in production for no more than 15 months during 1967 and 1968. It is a reflection of the US market's continuing importance to TR sales at this time that nearly three times as many TR250s were made as TR5s. In Britain, the TR's home country, the TR5 is quite a rare beast.

Left
Unlike that of the boot lid, the bonnet badging was plain and simple. The Triumph name had disappeared from the TR5's bonnet, and the only badge was an enamelled trapezoid offset to the left.

THE TESTERS' VIEW

It is normal evolution for an engine size to get bigger as a model is brought up to date at regular intervals. There was therefore nothing unusual about the TR capacity being increased from 2,138 c.c. to 2,498 c.c. last autumn, except that instead of the original lusty four-cylinder engine, a longer-stroke version of the Triumph 2000 six-cylinder unit was fitted.

Not content with the extra smoothness, refinement and a lot more torque, Standard Triumph decided to go for ultimate power and fitted Lucas petrol injection as standard equipment. From the TR4's 104 bhp *(78kW)* net at 4,700 rpm, the TR5's output goes up to 150 *(112kW)* at 5,500, so in terms of performance it is in a different class. Other mechanical changes include a beefed-up final drive with a 3.45 instead of 3.7-to-1 ratio, stiffer rear springs, a vacuum brake servo and bigger front discs. The gearbox is the same as before. Externally there are a few styling changes for identification of the model, the wheel trims being the most obvious, and inside there are revisions to comply with the US Federal safety regulations. The price, understandably, is £185 higher before tax.

Whatever else may be needed in the character of a sports car, the first essential is a lively engine and plenty of acceleration. There are no disappointments in the TR5 on this score, and its performance is a good match for many more powerful cars. Talking figures, the top speed of 120 mph *(193km/h)* (without overdrive) is 11 mph *(18km/h)* better than that of the overdrive TR4A we tested nearly two years ago. Accelerating through the gears from rest, 100 mph *(161km/h)* comes up in 28.5 sec, 20 sec earlier than with the TR4A. Anything under 17 sec for a standing quarter-mile *(0.4km)* is exciting; the TR5 does it in 16.8 sec, passing the post at 82 mph *(132km/h)* (TR4A 18.5 sec, 74 mph *[119km/h]*).

Although only of academic interest in such a car, by braking to below tick-over rpm, we found it possible to take top gear acceleration figures from 10 mph *(16km/h)* (470 rpm) whereas the TR4A (with twin Stromberg carburettors) would not pull reasonably below 20 mph *(32km/h)* (1,000 rpm). Apart from a slight "hiccup" between 12-14 mph *(19-23km/h)* (550-650 rpm), power is delivered smoothly and evenly throughout the range. Unfortunately one has to pay for this extra

Below
The spare wheel was normally concealed by a false boot floor, but extra space for soft baggage could be obtained by removing it.

urge, and the TR5 is the thirstiest TR so far. Our overall fuel consumption of 19.6 mpg *(7km/litre)* is of course partly the result of much hard driving, but it seems unlikely that even the most restrained behaviour would wring much more than 25 mpg *(9km/litre)* from it. Oil consumption at 350 miles *(563km)* per pint *(0.6 litre)* is slightly heavier than in the past.

In the morning it is essential to use the cold start knob (as Triumph call it). This shuts a butterfly in the induction air trunking and sets the metering unit to give excess fuel. No rich mixture, but a small amount of throttle opening is needed for hot starts. With the cold start device in use, the idling is smooth, but rather fast. If the knob is pushed back too soon, the engine fluffs and hesitates until warmed up, which it does very quickly. Without being offensively noisy, the TR5's twin exhaust pipes make an impressive and pleasingly smooth hum, rising to a full six-cylinder song when revved. Neither the gearbox nor final drive are audible. Wind noise is considerable at anything over 50 mph *(80km/h)*, especially

Left
From behind, the TR5 tell-tales were the plate badging on the boot lid, the reversing lights below the main lamp clusters and the dual-tipped exhaust pipe—here emphasized by chromed finishers, which were not standard. The rear wing badging read '2500', advertising the capacity of the new six-cylinder engine. The marker light was in red and had been added to meet new US lighting regulations. Instead of blanking off the hole in the wing on cars sold outside the USA, Triumph left the lamp in place.

Below
Wire wheels were optional, but not chrome examples as fitted here. The standard disc wheels were fitted with Rostyle full covers.

with the hood up so that the driver is certainly aware of speed.

One seems to be sitting very much in the cockpit, with the quite long high-sided bonnet stretching in front. Hard acceleration makes the back squat and the nose lift potently, though the actual change of attitude feels greater than it is. The engine responds immediately to the throttle with plenty of power always; from about 3,000 rpm on there is a firm shove in the back without a sudden jump in torque. The red area of the rev counter begins at 5,500 rpm but for our performance measurements we changed gear at 6,000 rpm; this gives maxima of 41, 64 and 96 mph *(66, 103 and 154km/h)* in the indirect gears.

Ratios are well chosen and there is powerful synchromesh. The gearchange is precise with a notchy action that sometimes makes it difficult to get into first at rest. Nevertheless it is delightful to use. The clutch is rather heavy by modern standards but it grips well, coping comfortably with 5,500 rpm wheel-spinning standing starts on MIRA's dry concrete, and a restart on the 1-in-3 test hill.

With 3.5 turns between 35½ ft *(10.8m)* locks, the TR5's rack and pinion steering is quick to respond to any movement, but the effort when manoeuvring is heavy. Feel is excellent without too much kickback on poor surfaces. Except in strong sidewinds on motorways, straight-ahead stability at speed is reassuring.

The test car was shod with Michelin asymmetric XAS radial-ply tyres which gave good grip in all conditions. Obviously with all that power available one has to tread carefully on the accelerator in the wet, otherwise the wheels spin in the lower gears and the tail snakes about. Except when rounding very sharp corners fast in first or second, it proved impossible to break the back away with power on well-surfaced dry roads. At MIRA we found that the car's naturally strong understeer grew progressively greater as cornering speed increased. The only way to tighten one's line under such extreme driving was to back off the accelerator (the natural reaction in an emergency), whereupon weight transfer forward and some small camber angle changes at the rear wheels cause the rear end to swing out.

Normally there is little body roll except when cornering really hard. The ride from the Triumph 2000 type all-independent suspension is good on ordinary smooth British roads, but not so good on rough ones. Sharp ridges and cats' eyes produce a noticeable bump-thump but the tyres themselves do not whine or roar on coarse noise-making surfaces. The only evidence of any body shake on rough stretches was an occasional release of the bonnet catch which was cured by resetting the latch plate on our test car.

One of the first things which owners of earlier TR models will notice is the much lighter servo-assisted braking. Maximum retardation of 1g is possible with the front wheels on the point of locking but it takes only 80 lb *(36kg)* pedal load now instead of the TR4A's 125 lb *(57kg)*. Our fade test from 70 mph *(113km/h)* caused the 0.5g pedal load to rise after 10 stops, from 45 to 55 lb *(20 to 25kg)*, which is still light. The handbrake is not the fly-off type; it gave 0.3g with both rear wheels locked and held the car on a 1-in-4 slope facing either way.

On the whole the driving position will suit most people well. The seats have a good wrap-round shape which locates one firmly during hard cornering. They are softly padded and very comfortable so that one gets out of the car after a long journey without feeling cramped or stiff. The backs are quite noticeably raked but do not recline. Padding over the whole steering wheel makes it look different from that of previous TRs, but the original four-wire spoke construction is still there behind the pvc. Every switch is identified with a symbol. The square flat knobs of the heater and mixture controls are shrouded by a crushable surrounding lip; those on our car were slightly twisted which upset those of us with a tidy eye. Two-speed wiper and the electric washer switches are the flush, rocker type mounted within fingertip reach of the steering wheel. Other safety points are the deep padding round the matt-finish wooden facia, the very clear black-bezelled, reflection-free instruments, rubber knobs on the window winders, near-flush door handles and hood locks which are cleverly, but not inaccessibly,

Left

The most significant improvement that the TR5 offered lay under the bonnet. Not only was a six-cylinder engine installed in place of the old four, but also fuel injection had replaced the SU carburettors. The TR5's new-found performance was more than welcome, as there had been little change in that department since the first TRs had gone on sale in 1953! With 150bhp (112kW) available, the TR5 was significantly faster than all its predecessors, and its 120mph (193km/h) top speed was a good 10mph (16km/h) higher than any of them could manage in standard tune.

Left

In the USA, Triumph was unable to offer the injected engine, so the company called the TR4A's replacement a TR250. It had the six-cylinder engine, but with carburettors, and Triumph tried to compensate by adding some garish stripes across its nose to make the car look as if it might be faster than the previous model. This example was re-imported to the UK in the late 1980s.

mounted on top of the windscreen frame out of the way of heads. Both the speedometer and rev counter proved accurate; twin horns are standard and give a clear penetrating note.

Foot controls are good in some ways and irritating in others. Heel-and-toes changes are easy but there is no proper place to rest one's clutch foot. What looks like a Phillips Stick-a-Sole stuck to the carpet on the sloping side of the transmission tunnel is a half-hearted effort more to prevent a hole appearing in the carpet than to provide somewhere other than the pedal as a rest. The dipswitch is up behind the clutch and hard to find.

Visibility is naturally best with the hood down, but with it up there are remarkably few blind areas, thanks to thin screen pillars and good use of the transparent panels. The high wing line fore and aft makes parking easy and only the wipers on our car were any sort of

obstruction, lying in front of the driver. Fitting of the flat-lensed outside mirror on the driver's door as required by US regulations is a help, but it needs a considerable deflection of one's eyes from straight ahead. A convex mirror on the end of the wing would be preferred by some drivers.

The heater is powerful but difficult to control progressively. There are three positions for the pull-out distribution controls: fully in—warm air from adjustable facia nozzles only; half way out (not easy to find)—hot air demisting, cold air for face; fully out—hot air for demisting and feet, cold air for face. A knob under the facia opens or shuts the scuttle intake for the heater air.

Luggage accommodation is reasonable for this sort of car without being exceptional. With the false floor over the spare wheel removed you gain a considerable volume for

Right
From the rear, there is nothing to distinguish the TR250 from the TR5, until you get close enough to read the badges. 'TR250' plates are fitted to the boot lid and rear wings.

Left
The dashboard mirrors that of the TR5, with padding along the top and around the switches to meet US safety specs. This example still has its original padded wheel.

Below
To meet US emissions standards, Triumph dispensed with fuel injection in favour of twin Stromberg carburettors.

Bottom
Despite the go-faster stripes, the TR250's carbureted six offered little extra performance than the TR4A's four-cylinder engine.

extra squashy baggage. In the cockpit there is a good locker in front of the passenger and even with the hood down (providing it is properly stowed) a lot of space in the trench behind the seats.

Folding the hood down takes a little longer than putting it up, because of the obstinate press-studs on the cover (two of which pulled out). Even so, it is still an exceptionally quick and easy single-handed job.

Driving the TR5 with the hood down is undoubtedly the best way to enjoy it to the full, even in winter. Rain is swept overhead at any speed over 40 mph *(64km/h)* and with the side windows up, buffeting at high speed is cut down slightly, though the square lines create a lot of back draught.

Maintenance and repair of relatively complicated fuel injection equipment is treated as a matter of course on diesel engines but the thought may alarm car owners who understandably prefer to do everything themselves. Triumph say that the only attention needed is renewal of the filter element (which lives behind a panel in the boot) at 12,000-mile *(19,312km)* intervals and a unit overhaul at 36,000 miles *(57,935km)*. Lucas tell us that experience with their Mk 1 system fitted to some production Maseratis for nearly 10 years

has shown that it is unlikely that replacement of metering unit, pump or injectors will be necessary during the life of the car. If any is, spares will be available on exchange. Tickover adjustment and re-synchronising of the three throttle spindles (one to each pair of inlets) after a decoke are quite straightforward. A clogged injector may be blown clear with an air-line. It is worth remembering when considering the merits of a petrol injection system that re-tuning of the previous twin-carburettor models was recommended at every 12,000 miles.

Headlamps give a reasonable light on full beam for up to 70 mph *(113km/h)*, but on a car of this performance iodine vapour main or auxiliary lamps would be appropriate. There are good automatic reversing lamps as standard, but no interior light of any kind.

Compared with its predecessors, the TR5 is a complete transformation. As we have remarked before, the heart of a car is its engine and the TR5 has an eager unit which responds just as a sports car driver wants. The rest of the car is traditional rather than dated, well modernised so that overall and above all it is very much a fun car still.

Reprinted from Autocar, 4 April 1968.

TRIUMPH TR5—SPECIFICATIONS

Engine	Six-cylinder, OHV, fuel-injected
Capacity	2498cc (74.7mm bore, 95mm stroke)
Max. power	150bhp (112kW) @ 5500rpm
Max. torque	164lb/ft (110kg/m) @ 3500rpm
Transmission	Four-speed, manual all-synchromesh gearbox with optional overdrive
Suspension, front	Independent, with coil springs, wishbones and telescopic dampers
Suspension, rear	Independent, with coil springs, semi-trailing arms and lever-arm dampers
Steering	Rack-and-pinion
Brakes	Discs on the front wheels and drums on the rear wheels; servo-assisted
Tyres	165x15 radial-ply
Length	12ft 9.6in (3.91m)
Width	4ft 10in (1.47m)
Height	4ft 2in (1.27m)
Wheelbase	7ft 4in (2.24m)
Max. speed	120mph (193km/h)
0-60mph (97km/h)	8.8sec
O'all fuel consumption	19.6mpg (7km/litre)
Production total	2947 (1161 for home market; 1786 for export)

TRIUMPH TR250—SPECIFICATIONS

Engine	Six-cylinder, OHV, carbureted
Capacity	2498cc (74.7mm bore, 95mm stroke)
Max. power	104bhp (78kW) @ 4500rpm
Max. torque	143lb/ft (213kg/m) @ 3000rpm
Transmission	Four-speed, manual all-synchromesh gearbox with optional overdrive
Suspension, front	Independent, with coil springs, wishbones and telescopic dampers
Suspension, rear	Independent, with coil springs, semi-trailing arms and lever-arm dampers
Steering	Rack-and-pinion
Brakes	Discs on the front wheels and drums on the rear wheels; servo-assisted
Tyres	185x15 radial-ply
Length	12ft 9.6in (3.91m)
Width	4ft 10in (1.47m)
Height	4ft 2in (1.27m)
Wheelbase	7ft 4in (2.24m)
Max. speed	107mph (172km/h)
0-60mph (97km/h)	10.6sec
O'all fuel consumption	27.5mpg (10km/litre)
Production total	8484 (all for US market)

TR6

Many people see the TR6 as the last of the great British sports cars and as the last 'real' TR. While this isn't the place to take sides in that argument, it is certainly true that the TR6 was the last of the old-style TRs, for its successors no longer had a separate chassis. It is also true that the TR6 was quite unrefined by the standards of its contemporaries. That was undoubtedly what gave it so much of its character, and that character helped to sell 94,619 examples in seven years—more than twice as many as any previous TR model.

Under its attractive new skin, the TR6 was a TR5 or TR250, with the 2.5-litre six-cylinder engine in fuel-injected (or, for the USA, carbureted) form in the IRS chassis. As regular styling consultant Michelotti was too busy to come up with the restyle Triumph wanted in the time available, the contract went to Karmann of Osnabrück, which was best-known for its work with Volkswagen, Porsche and BMW. But the finished product was not all Karmann's: Triumph incorporated a squared-off Kamm tail from an unrelated Michelotti prototype, and designed the new one-piece hardtop in-house.

In its seven years of production, the TR6 underwent no great changes. Production began in November 1968, although cars did not reach the showrooms until the following January. A number of small revisions followed that autumn, when the tacky Rostyle wheel

Facing page
Cars built in the 1970-72 model years had these silver painted wheels with black centres bearing a TR6 badge.

Below
On 1970-season and later cars, the windscreen pillars and top rail were painted matt black. This picture of a 1972 model demonstrates the TR6's purposeful lines.

Above
The magnificent under-bonnet view of a 150bhp (112kW)—pre-1973—TR6.

trims were dropped in favour of dished and perforated steel disc wheels, which contributed significantly to the car's distinctively brutal looks. Rake-adjustable seats arrived with better shaping; the steering wheel acquired satin-finish spokes; and a matt black windscreen surround was added to make the glass area appear larger.

The next change came in mid-1971, when the gearbox was swapped for the stronger Stag type, with wider-spaced ratios. A Laycock J-type overdrive replaced the old A-type in January 1973, and wire wheels ceased to be optional in May, but the major turning point in the TR6's production life occurred that

autumn. Most important of the 1974 season changes was a new camshaft, intended to improve the lumpy tick-over and low-speed running of the engine. It succeeded in that, but also took away some of the top-end power. The loss was not as great as figures suggest, however, because the 150bhp (112kW) of the original engine was calculated to SAE standards, while the 124bhp (92kW) of the 1974 season engine was calculated to the much stricter DIN standard.

Along with the revised engine came other changes. Minor instruments were made more legible with needles that pointed upwards instead of downwards, the dipswitch was moved to the steering column, and the steering wheel was restyled. Black wiper arms and blades were fitted, and wheel centre trims had a satin finish instead of being black. There was a new bib spoiler at the front, and the Triumph badge at the rear was relocated from the tail panel to the bumper.

The last TR6s for the home market were built in February 1975. Carbureted North American-specification cars, however, continued alongside the new TR7 until July 1976.

Like the TR4A, TR5 and TR250, the TR6 never found a place in the works competition teams. However, Bob Tullius successfully campaigned a car in SCCA racing in the USA.

Left
This publicity shot taken for the Swiss market shows the TR6's squared-off Kamm tail with its matt black rear panel, and the white TR6 decal at the rear of the wings. On light-coloured cars, this decal was black. The Rostyle wheel trims and body-colour windscreen pillars were available only on 1969-season cars like this one.

THE TESTERS' VIEW

Facing page, top
TR6s had an identifying badge in the middle of the grille. This is the early type of grille, without the chrome strips at top and bottom, which were added for the 1973-model year.

Facing page, bottom
All models of TR6 had three-piece rear bumpers and double-barrelled exhausts. After the 1972-model year, to which this car belongs, the number-plate lights were moved from their bumper mounted plinth to the lower edge of the boot lid.

Even if the Austin-Healey 3000 had not been dropped, the TR6 would have taken over as the he-man's sports car in its own right. It is very much a masculine machine, calling for beefy muscles, bold decisions and even ruthlessness on occasions. It could be dubbed the last of the real sports cars, because it displays many qualities so beloved in vintage times. In spite of all this (although many would say because), it is a tremendously exhilarating car to drive anywhere.

To recap on the car's introduction, it was essentially a face-lift on the TR5, announced in January this year. The TR5 itself was too much like the TR4A and TR4 to catch on in the USA where there are an awful lot of Triumph sales. So Karmann Ghia were consulted for a quick rejuvenating process, and the result is a kind of smoothed out TR with fashionable and effective use of matt black for the radiator grille and undercut tail panel.

Mechanically the engine remains the same, with 2½ litres of six-cylinder power turning out 142 bhp *(106kW)* net with the aid of Lucas

fuel injection. (Cars for the USA have Stromberg Duplex carburettors to satisfy the Federal anti-smog regulations which reduces the peak of power curve only to 126 bhp *[94kW]*). There is the same independent rear suspension as on the TR4A, employing semi-trailing arms and coil springs.

It is necessary to use the mixture enrichening knob to start the engine in the morning, or if the car has been standing for more than about 20 minutes. As soon as possible, though, this control should be pushed back because to judge by the black smoke from the exhaust it overdoes things a bit for the British conditions. There are multi-lingual instructions on the windscreen about starting techniques, and by following these we had no trouble during the whole of our test. In general the fuel injection engine takes an extra turn on the key to start it, compared with the carburettor version.

As soon as it fires though, the advantages can be felt. It is a sweet and zippy unit, with an extremely sporty twang to its exhaust note

Right
There were no model identification badges on the rear panels of the TR6, just this discreet cast badge as a reminder that the engine was equipped with fuel injection.

Right
The TR6 initially had its chassis number plate attached to the front inner wing, not to the bulkhead, as had the earlier TRs...

Facing page, top
...while the later chassis number plate was located on the door pillar.

and real punch all the way through. There is a crankshaft vibration resonance at 6,000 rpm, so the rev counter scale stops at 5,500 with a big red sector from here on. It looks pretty emphatic, so we changed gear at the beginning of it, allowing a small margin for the slight error we measured in the instrument (reading 5,400 at true 5,500 rpm).

Acceleration

From rest to 30 mph *(48km/h)* takes only 2.8sec, and 60 mph *(97km/h)* comes up in a surging 8.2sec. It is possible to reach 100 mph *(161km/h)* in only 29sec from a standing start, the distance needed being about half a mile *(0.8km)* only.

These figures are very rapid for a 2½-litre sports car and a fraction or two better than those for the TR5. We felt that the engine on the test car had more torque low down than we

remembered on the TR5, and this could be the result of some small changes made recently to improve idling. The long-stroke engine still ticks over lumpily, but adjustment of the idling revs is now much simpler.

Whether the optional overdrive is fitted or not, the axle ratio is 3.45 to 1, giving a top gear speed of 21.2 mph *(34km/h)* per 1,000 rpm. This is a quiet and restful ratio for motorway cruising (3,300 rpm at 70 mph *[113km/h]*) and the car goes faster in direct than overdrive on level ground.

But as well as working on top, overdrive operates on second and third, so by flicking the steering column stalk one has almost the effect of a two-speed axle with town and motorway ranges. Overdrive second has a very useful maximum of 71 mph *(114km/h)*, while overdrive third is pretty well the same as top (1.08 instead of 1.0 to 1). Overdrive top

Left

Another publicity shot, this time of a 1973 model with that year's revisions. Most obvious are the under-bumper spoiler, silver wheel centres and black (instead of satin-finish chrome) wiper arms. Less visible is the bright strip across the bottom of the grille; a similar piece of trim ran across the top. This car is also fitted with the optional hardtop, which had become a one-piece affair; the 'Surrey' top was not offered for the TR6. Note also the small British Leyland badge at the trailing edge of the front wing. Expert opinion holds that this was fitted only on 1973-model and later cars, but factory publicity photographs suggest that it was fitted to some cars as early as the autumn of 1970.

95

is really meant for cruising, although it is surprising how much it can be used in town.

The engine is flexible enough to pull from 20 mph *(32km/h)* in direct top, although it is hardly likely that a TR driver would ever feel lazy enough to try it. Engagement and disengagement of the overdrive are both outstandingly smooth, which encourages frequent use of the unit.

We used every ratio available in the combined seven-speed gearbox when taking acceleration figures, making a split shift from overdrive second to direct third, and then back to overdrive at 88 mph *(142km/h)*. This saved about half a second only on using the simple four-speed manual gearbox, which only goes to show how well the ratios are matched to engine torque.

The gearbox has very effective synchromesh on all forward ratios, and it withstood the most violent snatch changes we could achieve when battling for every tenth on the MIRA test strip. The clutch took full power without slip during standing start runs, but it

was extremely heavy to operate. We cannot remember when, if ever, we last had a car which sent our pressometer needle so far round the dial when measuring the effort to free the clutch. These days anything over 30 lb *(14kg)* is beginning to qualify as heavy and the TR6 needed 67 lb *(30kg)*. On the TR5 the effort was only 35 lb *(16kg)*, so there may have been a fault particular to the test car.

Despite servo assistance, the brakes are heavy too, calling for over 100 lb *(45kg)* effort for an emergency stop. Mostly though they react with a reassuring bite, pulling the speed down very rapidly when required. Without any wheel locking or slewing we achieved over 1g, thanks largely to the Michelin XAS radial tyres which are standard. When tested for fade from 70 mph *(113km/h)*, the brakes remained remarkably stable with a slight change in their speed sensitivity warning the driver that they were getting hot.

The handbrake could not hold the TR6 on a 1 in 3 gradient, and it took a mighty heave to secure it on a 1 in 4. Restarting was managed

Below
The cockpit of a 1972 TR6. The satin-finish wood veneer dashboard was inherited from the TR5/TR250, but the TR6's seats had pleats running from side to side for the first time since the early 1960s.

only by carefully slipping the clutch to the top of the rise.

Steering and Roadholding

For a sports car which needs to be driven briskly through tight turns or in and out of traffic, the steering is too low geared and too slow to respond. The wheel feels larger than it is because it is hard for the driver to get far enough from it, and it takes almost 3½ turns between 36ft *(11m)* locks. On the road the car steers stably and holds its line well, but there is no immediate reaction to wheel movements and it has a dead feel out of character with the rest of the car.

Distinctly different in source, but much the same in effect, is the handling which displays predominant understeer in all conditions. Extending ourselves and the car to the limits of adhesion on the MIRA road circuit we eventually understeered almost straight ahead after applying full power in second on the apex of a dry corner. Unless traction is broken at the rear wheels (impossible in the dry) the

tail always stays in line, and even on wet surfaces there are no real problems from having so much power on tap. It would be better if the throttle linkage did not give quite such a snappy initial opening, but otherwise the TR6 handles safely at all times.

It is inevitable that a sports car with a fabric roof should suffer from an excess of wind noise and the TR6 is no exception. It is better than most in this respect, but not a restful car on long journeys. Mechanically it is very smooth and the miles disappear at a satisfying rate when touring.

Unfortunately sports car stability can only come, in this case, with a harsh and very firm ride. Even on the smoothest surfaces the TR6 feels very taut and joggly; on rough roads it thumps along and patters round corners with lots of action but little insulation for the occupants. In return for this there is virtually no roll on corners and no pitch. When applying full power the tail squats down most noticeably and the headlamp beams at night swing up to dazzle oncoming drivers.

Below
This is a pre-1973 US-market TR6, fitted with air conditioning, as were so many cars for that market. This example has the steering wheel with slotted metal spokes that was first seen on 1970 models.

We were very lucky with the weather for our TR test, and for much of the time we were able to stow the hood behind the seats (a simple and quick job once the rear window has been unzipped) and hide it under its neat pvc cover. In open trim the car really comes into its own with fresh air on all sides and a more audible exhaust note. Erecting the hood is really fast, the only delay being to locate the pegs in the screen rail accurately. Their steep-ramped helical locks take care of the tensioning automatically.

With an open car it is essential to have a heater powerful enough to cope with the extra air spillage when the roof is off. The TR6 unit puts a nice hot blast round one's feet on a cold morning, yet is reasonably controllable when the roof cuts down the cabin size. The temperature valve is operated by a stiff push-pull knob which ought by now to have been replaced by a smoother quadrant slide.

All knobs are easy to find and reach, some being rockers (washers and wipers) and some push-pullers (heater fan, clock). All the lamps are worked by a steering column stalk which can be pulled to flash main beams. The dip-switch is on the floor. At each end of the facia is a swivelling eyeball nozzle for fresh air.

Instruments are the best there are with large dials, clear lettering and smart matt-black bezels. Speedometer error was only slight (over-reading by 3 mph *[5km/h]* at 70 mph *[113km/h]*) and the mileage recorder (trip and total provided) showed 2.6 per cent more distance than the truth.

Seats have fixed backrests and soft upholstery with plenty of wrap-round. By current standards the cockpit is rather narrow leaving no room for armrests. All the inside is well trimmed and a matt imitation wood veneer facia looks luxurious and practical.

Despite the points we have criticised, we all enjoyed driving the TR6 and reckon it to be one of the best fun cars around right now. Like a heavily built hunter, it responds better to a firm hand and takes a man to get the best from it. In a drag match there are few cars which can stay with it. To go much quicker is going to cost several hundred pounds more at least, so for performance alone the price is a competitive one. For the few joys of motoring left, it is a bargain.

Reprinted from Autocar, 17 April 1969.

Right
1973-model and later TR6s sold in the USA had this patriotic decal applied to the rear wings.

Left
*From 1973, US-market TR6s
wore ungainly rubber
overriders on both front and
rear bumpers. Note also the
all-orange lamps above the
bumpers, which met US
regulations concerning
marker lights.*

TRIUMPH TR6—SPECIFICATIONS

Engine Six-cylinder, OHV, fuel-injected (carbureted for USA)

Capacity 2498cc (74.7mm bore, 95mm stroke)

Max. power 1970-72 fuel-injected models, 150bhp (112kW) @ 5500rpm;
1973-on fuel-injected models, 124bhp (92kW) DIN @ 5000rpm;
1970-71 carbureted models, 104bhp (78kW) @ 4500rpm; 1972-on
carbureted models, 106bhp (79kW) @ 4900rpm

Max. torque 1970-72 fuel-injected models, 164lb/ft (110kg/m) @ 3500rpm;
1973-on fuel-injected models, 143lb/ft (96kg/m) DIN @ 3500rpm;
1970-71 carbureted models, 143lb/ft (96kg/m) @ 3000rpm;
1972-on carbureted models, 133lb/ft (89kg/m) @ 3000rpm

Transmission Four-speed, manual all-synchromesh gearbox with optional
overdrive on top three gears (to end of 1972), or third and top
only (from January 1973); overdrive standard after December
1973

Suspension, front Independent, with coil springs, wishbones, anti-roll bar and
telescopic dampers

Suspension, rear Independent, with coil springs, semi-trailing arms and lever-arm
dampers

Steering Rack-and-pinion

Brakes Discs on the front wheels and drums on the rear wheels; servo-
assisted

Tyres 165x15 radial-ply

Length 13ft 3in (4.04m); 1973-74 US models, 13ft 6.1in (4.12m);
1975-76 US models 13ft 7.6in (4.16m)

Width 4ft 10in (1.47m)

Height 4ft 2in (1.27m) (with soft top erected)

Wheelbase 7ft 4in (2.24m)

Max. speed Fuel-injected models, 119mph (192km/h); carbureted models,
109mph (175km/h)

0-60mph (97km/h) Fuel-injected models, 8.2sec; carbureted models, 10.7sec

O'all fuel consumption Fuel-injected models, 19.8mpg (7km/litre); carbureted models,
25.9mpg (9km/litre)

Production total 94,619 (8370 for home market; 86,249 for export)

TR7

The TR7 upset a lot of traditional TR fans when it was announced in 1975. This wasn't just because it had reverted to a four-cylinder engine after the six-cylinder TR5/TR250 and TR6; nor was it just because it had modern monocoque construction instead of the traditional separate chassis and body. These factors undoubtedly had their effect, but the real 'crime' for which die-hard TR fans could never forgive the TR7 was that it was not an open car.

In fact, the TR7 had been designed as a closed coupé because Triumph (along with many other manufacturers) believed that proposed US legislation was likely to ban the sale in that country of open cars on safety grounds. As the US market was vital to TR sales, Triumph played for safety when designing their new model. As it turned out, the proposed laws had been defeated by the time the TR7 reached the showrooms, but it would be four more years before a redesigned, open TR7 would go into production.

By the time the TR7 was being planned, Triumph had lost a great deal of its independence within British Leyland, and the new car was intended to draw on expertise from all of BL's constituent parts rather than simply on Triumph's. Thus, although its slant-four 2-litre engine was based on the unit in Triumph's Dolomite saloons, the five-speed gearbox and

Facing page
One of the many distinctive features of the TR7 was its retractable headlights.

Below
This picture of an early right-hand-drive TR7 shows the 'wedge' shape to its best advantage; from some angles, the short wheelbase made the car look rather dumpy. Note the black wheel centre panels.

heavy-duty rear axle that became optional (and standard in the US) in 1976 were Rover items. Similarly, its controversial 'wedge' styling was drawn up in the Austin styling studios by Harris Mann.

The TR7 was built at the new Triumph factory in Speke, near Liverpool, a plant that was plagued with labour relations problems. The quality of many of the early cars was abysmal in consequence, and did nothing for the TR7's reputation. Quality improved only when production was transferred to the Canley factory in 1978, after a four-month strike at Speke had halted the TR7 altogether. For its final year, the TR7 was built in the Rover plant at Solihull.

However, the TR7 was by no means a bad car, and today it has a strong enthusiast following. It was also by far the most successful of all the TRs, outselling even the TR6 by some 18,000, most of which swelled sales in the home market. In the export markets, which really counted, however, it sold more slowly than the TR6, and BL terminated production just as public perceptions of the car were becoming more positive after the introduction of the drophead model.

With the TR7, BL also re-introduced TRs to competition in works colours, and the cars did quite well in European rallies in the mid-1970s, before being replaced by TR7-V8s (effectively pre-production TR8s) in the works team. In the USA, John Buffum also won his SCCA class in a TR7 during 1978.

There was a wide variety of different specifications for the TR7. US-market models came in very different states of tune from those for the rest of the world, and even within the USA there were different engine specifications for California and for the other 49 states. The final US models were equipped with fuel injection instead of the carburettors universally fitted for other markets. In addition, both the UK and US markets were offered a number of limited-edition models (with special paint, equipment, or a combination of these). Adding to the variety were automatic-transmission models, for the TR7 was also the first TR to be offered with this feature.

Below
From dead ahead, the TR7 also had an attractive shape. The large under-bumper air intake lip is clearly visible in this shot of an early left-hand-drive car.

All cars built at the Speke factory had a large TR7 decal on the nose, as on this well-preserved example.

Below
From behind, the same well-kept car shows the rear badging used on the Speke-built TR7s. The tail panel on these cars was given a matt black finish.

THE TESTERS' VIEW

The Triumph TR7 was designed for America, and that explains a lot. It explains why it has a maximum speed of only 109 mph *(175km/h)* and acceleration that is no more than moderate in its class. Speed limits even lower than those of increasingly restrictive Europe may also explain why the gearing is such that the TR7 is uncomfortably fussy when cruising on a motorway at much over 70 mph *(113km/h)*. Competition in the US marketplace dictated the well-planned and well-equipped interior and the eye-catching if stylized appearance. America also explains why the latest in a proud line of true sports cars has broken with what many would consider to be the most important tradition of all and is offered only in fixed-roof form; US roll-over regulations proposed at the time of its design, but not subsequently implemented, seemed to signal the end of the open sports car.

It is important to establish at the outset that the TR7 is not a direct descendant of the TR line. It has virtually nothing in common with its predecessor, the TR6. Gone are the smooth but punchy six-cylinder engine, the fuel injection, the overdrive, the independent rear suspension. It is better to think of the TR7 as a two-seater version of the four-cylinder Dolomite with MacPherson strut front suspension instead of double wishbones. Viewed in that way its performance is entirely satisfactory with acceleration and maximum speed closer to the 16-valve 2-litre Dolomite Sprint than the ordinary Dolomite 1850, but with overall fuel consumption approaching that of the latter. Viewed as a gap-filler for Leyland, the TR7 has better all-round performance than the MGB (which continues) and the Triumph GT6 (which ceased production in 1974 and is in many ways the TR7's most direct ancestor).

Right
Built at the Solihull factory, this Bordeaux Red car is a good example of the later fixed-head coupé, which had a Webasto sliding roof as standard equipment.

Today's real sports cars, at least in the more mundane price classes, are saloons like the Dolomite Sprint and the Ford RS2000. The TR7's closest competitor on performance is the Toyota Celica GT, no longer sold in Britain. Without performance advantages—and the possibility of putting the roof down—the modern two-seater has to be able to provide something else that sets it apart from other cars. Hence, on the one hand the TR7's controversial styling (by Harris Mann, who also designed the Princess saloon), and on the other its exciting interior. The latter seems to have taken a leaf out of the Lancia Stratos book. One climbs down into an inviting cockpit, is enclosed by curved doors and a steeply raked screen, confronted by a massive facia complex, and accommodated in well-designed and well-sited seats. The side and rear windows are quite shallow, and the facia top and rear shelf are high so that occupants have the impression of sitting low down enveloped by the bodywork—this, not the

Above
The streamlined nose of the TR7 left no room for conventional headlamps, so they were fitted into motor driven pods that rose when the lights were needed.

flapping side-screen wind-in-the-hair convertible, is the modern idiom of the sports car. Psychologically, Leyland's designers have done well; the way the TR7 turns heads suggests that whatever one may think of the styling, they have got the "image" right.

Leyland have also got the price right; at 12 pence less than £3,000 it represents good value by today's inflated standards, and is pitched neatly between the MGB GT, Dolomite 1850 and the TR6's last quoted price (all around £250 cheaper) and the Dolomite Sprint and the competitors from Lancia, BMW and others.

Price considerations played a part in the decision to introduce the TR7 with a 2-litre version of the 8-valve Dolomite single overhead camshaft engine rather than the more powerful 16-valve of the Dolomite Sprint, which is used in the factory-prepared TR7 rally cars. Bore and stroke are therefore the same as the Sprint at 90.3 mm and 78 mm respectively, giving an actual capacity of 1,998 c.c. Compression ratio is between the Dolomite 1850 and the Sprint at 9.25 to 1 and

significantly higher than that of the heavily emissions-controlled version for North America. Four-star fuel is recommended; the tank carries 12 gallons *(55 litres)*.

A maximum of 105 bhp *(78kW)* at 5,500 rpm also puts the TR7 engine squarely between the 8 and 16-valve Dolomites. Unlike the American version, which has Stromberg carburettors to go with a variety of anti-pollution devices, British-specification TR7s use twin SU HS6 carburettors. They do not have Lucas electronic ignition which is a feature of the Federalized cars.

The gearbox is the familiar single-rail unit which has found its way into a number of rear-drive Leyland cars. The ratios are the same as the Dolomite 1850 and so is the 3.63 to 1 final drive, but with the important difference that the TR7 is not available with the Dolomite's overdrive. With that omission the "ideal gearing" referred to in our last Dolomite Autotest cannot apply to the TR7. Only 17.9 mph *(29km/h)* per 1,000 rpm in top is too low for relaxed cruising, particularly since the test car at least had a harshness around 4,000 rpm

Below

Inside the TR7, the basic dashboard layout never changed, but the broadcord upholstery seen in this early car did not last beyond the first few months of 1977.

Left
Apertures for the side marker lights demanded in the US market were filled with black plastic blanking plates on cars sold elsewhere. This is the plate fitted at the front.

Below
When production was transferred to Canley in 1978, the cars were heavily revised in detail. Canley-built TR7s can be identified by the wreath badge on the nose panel. Note the amber side marker lamp on this 1980-model British registered car, actually a US-market item.

which coincides with the legal motorway cruising speed.

We have previously suggested that the 8-valve 1,850 c.c. Dolomite engine has lots of bottom end and no top and the TR7's performance shows that this still applies, though to a lesser extent to the full 2-litre unit. So, like the saloon, the engine is usefully flexible and therefore easy and economical to drive at traffic speeds. It also allows the use of an unusually high first gear with a maximum of 44 mph *(71km/h)* at the 6,500 rpm rev-counter red line, though the noise level (and the fact that this is 1,000 rpm over the power peak) do not encourage such high revving.

The test car proved an easy starter, with little choke needed during the mild weather of the test, and an even 500 rpm tick-over (though after a lot of high speed running, some stiction in the throttle cable tended to leave a much faster idle). The clutch screeched in an alarming but inconsequential manner after a morning cold start. Driveaway characteristics are good and the engine is unobtrusive at low speeds, with more noise

(whine) being generated by the gearbox than the power unit. However, a hammering sort of roughness arises at around 4,000 rpm and continues until beyond 5,000 rpm, which coincides with the most useful part of the power band from both a performance and top gear touring point-of-view. Through this critical 1,000 rpm period the roughness seems to be a combination of engine noise, body/chassis boom and vibration transmitted through the gearbox. At a steady 80 mph *(129km/h)* (4,500 rpm) a heterodyning drone is detectable. It should be said that our experience with other TR7s suggests that while all have some harshness and high noise/vibration periods, when and how these appear are not consistent. It is also worth pointing out that in the test car, the passenger heard all this rather less clearly than the driver.

But if the need for an overdrive is emphasised by the noise, the poor change quality of the existing gearbox (it is notably reluctant to engage both first and reverse gears) makes it clear that the TR7 badly needs this weakest link replaced by the excellent five-speed box

Left
*The press were full of praise
for the spacious interior of
the TR7. They liked the clear
instrumentation and
comfortable seats.*

Left
*A fine example of a 1981
drophead TR7. With the top
down, it was an impressive
looking car.*

of the forthcoming new Rover. It is already used in the works rally cars and reference to it in the TR7 handbook suggests that this option will soon be available. The ratios listed (with a 3.9 to 1 final drive) are the same as in the Rover and will give a much more useful 21 mph *(34km/h)* per 1,000 rpm in top.

The present arrangement does provide reasonable economy. A 2,000 mile round trip to the South of France with a high proportion of motorway cruising at around 85 mph *(137km/h)* produced an overall consumption figure of 27.6 mpg *(10km/litre)*, while a hard-driven period of test driving in this country, including taking the performance figures at the MIRA Proving Ground, returned 23.7 mpg *(8km/litre)*. At a constant 70 mph *(113km/h)* the TR7 achieved 29.6 mpg *(10km/litre)*; a flat-out lap of the MIRA banked circuit gave 16.5 mpg *(6km/litre)*.

The TR7's maximum speed of 109 mph *(175km/h)* is slightly lower than Leyland's claim but the 9.1 sec 0-60 mph *(97km/h)* time is rather better than they led us to expect. The argument goes that with most of the world covered in speed limits the maximum is "enough"; be that as it may, the TR7 will not happily cruise at over 100 mph *(161km/h)* like its predecessors.

Much better handling

Whatever else the old TRs might have done, they didn't, in the final analysis, go round corners all that well. Triumph's thoroughly modern TR7 may have an old-fashioned live axle but it beats its predecessors hands down when it comes to handling and roadholding. That it does so convincingly is a tribute to the development of a very simple and well-tried suspension system. Triumph's advertisements in the United States make much of this engineering simplicity. That refers to the adoption of MacPherson strut front suspension, using the anti-roll bar as a bottom link, and the live axle on four links with an anti-roll bar attached—unusually—to the bottom trailing arms. Close attention has been paid to spring rates and damping, the philosophy being that a good (read, well developed) live axle is better (and cheaper) than a mediocre independent rear suspension system.

The suspension, front and rear, has long travel. The TR7 is designed to keep off its bump stops whereas too many low-slung sports cars spend too much time using this inconsistent additional springing medium, with correspondingly unfortunate effects on their handling. The result is that the TR7 soaks up uneven surfaces very well indeed. It can be driven fast over a series of conflicting bumps—a level crossing in the middle of an S-bend for example—with great confidence. Bumps produce very little steering deflection and there is an inherently low degree of body roll. That all goes to make the TR7 a fast and satisfying car to drive on give-and-take country roads—and a comfortable one to travel in. In this sense, the TR7 is a long way from the rough ride standards of old-fashioned sports cars and we think that this is a tradition that deserved to be broken.

The steering is sweeter than in some of the TR7's hairy-chested predecessors—precise, responsive, moderately geared and requiring only moderate turning effort. The car's basic

Facing page, top left and right
Outside the USA, TR7s had twin SU carburettors. There was plenty of room around the 2-litre engine because the engine bay had been designed from the start to accommodate the physically larger 3.5-litre aluminium Rover V8 engine.

Facing page, bottom
Cars built at Canley and Solihull had different rear badging to the models from Speke, and a tail panel in the body colour.

Right

*For the 1980-model year,
smart alloy wheels became
standard on cars destined for
the USA, and optional on
those sold elsewhere.*

handling characteristic is understeer, which it exhibits only slightly on long, fast motorway bends and unobtrusively at slower speeds. That makes for reasonable straight-line stability. Rear end breakaway is not difficult to control. Roadholding is generally good, in both dry and wet conditions. The standard tyre wear is 175/60 SR 13 Goodyear G800S steel braced low-profile radials. Interestingly, the handbook describes the tyre fitment for the so far unannounced five-speed version as the wider 185 tyres though still of SR speed rating which applies only up to 113 mph *(182km/h).*

The only thing that upsets the TR7's generally impressive road behaviour is the brakes. They do not inspire confidence on first acquaintance because the servo-assisted pedal is spongy, a characteristic shared with some Dolomites we have driven. This means that there is some lost travel before even gentle check braking can be achieved. Pushed harder, the bite is more reassuring, though heavy braking from high speeds produces a rumbling near to the final stopping point. A fast downhill run in the French Alps produced a lot of smoke from the front brakes, and a little fade, which our MIRA tests were unable to reproduce. More disturbing than this is the haphazard way in which the wheels lock in an emergency situation. Our braking tests demonstrated that the front wheels always lock before the rear but which front locks first varies; the early point at which wheel lock occurred also accounts for an optimum reading of 0.85g which is not very good by modern standards. The braking system has a rear pressure-limiting valve, but in the test car it was rather over-enthusiastic. The uncertain feeling is, not surprisingly, even more marked in the wet and the greatest inhibitor to anything like fast driving on slippery surfaces, which in other ways the TR7 negotiates well.

Interior comfort

The interior is not the usual cramped and spartan sports car style, though the styling has dictated that the seating position is so far back that the seat recliners serve only to fine-tune the driving position; the seat bases can be pushed back to the rear bulkhead. The seats themselves are among the best-shaped we have encountered in a sporting car of any price. Covered in corduroy-type material they have flat looking but well moulded cushions and dished backs; adjustable head restraints

Right

*For the final year of
production, TR7s were built at
the Rover factory in Solihull.
Although early cars had
serious build-quality
problems, the Solihull models
were very good indeed.
However, the damage to the
TR7's reputation had been
done, and production was
halted in 1981.*

Left
*Cars built at Solihull for the
1981-model year had a
distinctive badge on the nose.*

are built in. The range of adjustment and the relative position of pedals and thick rimmed trim-covered steering wheel are such that all our testers gave the driving position high marks. Also praiseworthy is the seat belt installation. The inertia reels are hidden behind the rear trim in the fuel tank compartment and the central buckles are mounted to the seat runners so that they remain in the correct position however the seats are adjusted; an example other manufacturers would do well to follow.

Not so good is the visibility all round from the cockpit. Here the styling gets in the way again for the nose position is totally out of sight (is that why it needs such substantial bumpers?) and over-the-shoulder rear view is obstructed by the heavy quarter panels. One sits low in the TR7, which is (by sports car standards) a wide car, and looks over a massive facia moulding to a steeply raked screen. The effect is that the view forward, though adequate, is shallow; to the rear, the close, vertical rear window is fine. The wipers, pantographic on the driving side, clear the big screen efficiently.

That massive facia contains clear instru-

ments and well-placed controls. The instruments, white lettering on black, are enclosed by a single sheet of glass, BMW-style. The speedometer and rev counter are matched either side of a vertical row of warning lights while the smaller dials are a fuel gauge, temperature gauge, voltmeter, and clock; lack of an oil pressure gauge is a notable omission for a car of this type.

Switches on the central panel above the main fresh air ducts operate the lights—which pop up electrically and very quickly (0.8 sec)—the heated rear window, and the hazard flashers, while the excellent Leyland design combination column stalks look after wiping/washing and dipping/indicating/flashing and beeping operations. Below the fresh air vents are the four vertical slide controls of the neat new BL-corporate air blending heater and ventilating system. Simple and effective, the levers control ventilators, temperature distribution and fresh air intake; the latter needing the lowest of its three fan speeds for adequate air-flow most of the time. Incidentally, opening the right hand side window at speed results in little buffeting for the driver, though there is a moderate amount of wind noise at

Right
This view of Alan Thomas' drophead shows that the soft top did not detract from the TR7's dramatic frontal appearance. The car is an early home-market example, built at Canley.

Left
*The TR7's convertible top
looked neat from all angles,
and did credit to those who
designed it.*

Left
*When stowed, the soft top of
the drophead was concealed
under a neat hood cover. The
car looked superb from
behind with the top down.*

high speed when closed. The TR7 handbook makes reference to an optional opening roof which was revealed as a canvas roll-back arrangement at the time of the US-specification TR7 announcement last year but has yet to be made available. The roof itself is pressed in such a way that sunroof installation should be easy.

Since the rear bulkhead is immediately behind the seats the TR7 has as little interior stowage space as many mid-engined cars. If one is prepared to have the passenger seat pushed well forward there is room for a briefcase behind, but a shopping basket would be difficult to stow inside, not to mention more bulky packages. There are lots of compartments for small items including three promising mouldings on the facia top which are virtually useless because of the reflections anything carried there creates, and some deeper ones on the rear shelf which can carry books etc as well as a jacket across the top. There is a reasonably sized metal-lidded glove locker, a narrow central cubby behind the handbrake and two tiny map pockets inside the rear flanks. The interior of the test car was all black and the fit of the various trim mouldings and carpet not of a high standard.

Though stowage space inside is somewhat restricted, there is, thanks to the width, the impression of plenty of space for the occu-

pants. The width also enhances the useful luggage boot which is just deep enough to carry two small suitcases on top of one another and wide enough to carry a number of soft bags; big enough for the holiday luggage of an average couple. The spare wheel and jack live beneath a false floor in the boot, under a poorly fitting mat. The tail panel carries the retractable radio aerial, which like loudspeakers in the doors, comes as standard though a radio does not.

In conclusion

Accommodation and comfort are two of the TR7's strong points; another is the enjoyable way it negotiates twisting roads. It is, for Triumph at least, a new sort of two-seater which may appeal to those who felt that such cars were too stark and uncomfortable for their needs.

The TR7 is a one-model range at present and there is yet no sign of when more potent or more highly-geared versions will become available, or indeed of exactly what form those will take. Until they appear, the Dolomite Sprint remains the real sports car of the Triumph range but the export-come-home TR7 represents an interesting, eye-catching, and entertaining alternative.

Reprinted from Autocar, 26 June 1976.

TRIUMPH TR7—SPECIFICATIONS

Engine	Four-cylinder, OHC, carbureted; 1981-season US models, fuel-injected
Capacity	1998cc (90.3mm bore, 78mm stroke)
Max. power	105bhp (78kW) DIN @ 5500rpm; US federal models, 92bhp (69kW) DIN @ 5000rpm
Max. torque	119lb/ft (177kg/m) @ 3500rpm; US federal models, 115lb/ft (171kg.m) @ 3500rpm
Transmission	Four-speed, manual all-synchromesh gearbox; optional three-speed automatic. A five-speed manual replaced the four-speed in US models for 1976, and was optional for other markets for 1976-77; it was standard for all markets for 1979-81
Suspension, front	Independent, with coil springs, MacPherson struts, anti-roll bar and telescopic dampers
Suspension, rear	Live axle, with coil springs, radius arms, anti-roll bar and telescopic dampers
Steering	Rack-and-pinion
Brakes	Discs on the front wheels and drums on the rear wheels; servo-assisted
Tyres	185/70x13 or 175x13 radial-ply
Length	13ft 4.1in (4.07m); US federal models, 13ft 8.5in (4.18m)
Width	5ft 6.2in (1.61m)
Height	4ft 1.9in (1.26m)
Wheelbase	7ft 1in (2.16m)
Max. speed	109mph (175km/h)
0-60mph (97km/h)	9.1sec; US federal models, 11.3sec
O'all fuel consumption	26.4mpg (9km/litre); US federal models, 34.4mpg (12km/litre)
Production total	112,375 (26,796 for home market; 85,579 for export)

TR8

The TR8 was planned as a companion model to the TR7 right from the beginning of the TR7 project, and the two cars were developed in tandem during the early 1970s. Like the TR7, the TR8 drew on expertise from outside the Triumph area of BL, and was planned to have the Rover 3.5-litre V8 engine. This, with 155bhp (116kW) on tap, would have made the TR8 the real successor to the 150bhp (112kW) six-cylinder TR6 if things had gone according to plan.

The TR8 should have become available in 1978, initially for the US market, but its production was halted by the strike at Speke. When it did enter production, with the drophead body and in emissions controlled US form for the 1980 season, it was extremely well received. A more powerful version, without the emissions control gear, was being prepared for other markets in 1981 when British Leyland decided to stop production of the TR7 and TR8 altogether.

Both in the USA and in Britain, where the few right-hand-drive evaluation cars were eventually sold off, the TR8 is highly prized by enthusiasts. It offers exciting performance together with the high degree of refinement associated with the TR7. A tribute to its desirability is that, in Britain, there is a thriving industry devoted to the conversion of TR7s to TR8

Facing page
This emissions controlled fuel-injected engine is in a 1981 TR8 drophead. Originally, it was a left-hand-drive example, but it has been converted to right-hand-drive. Tough emissions legislation in California demanded fuel injection for the TR8, when cars intended for other markets had carburettors.

Below
Other than discreet badging, there is little to differentiate a TR8 from a TR7.

specification, often with additional performance enhancements.

The TR8 did appear as a works rally car in 1978, when it was known as the TR7-V8 because at that stage the TR8 itself had not gone on sale. Between then and 1980, the works TR7-V8s racked up an impressive number of rally victories, but BL decided to withdraw from competition for financial reasons and the cars never really achieved their full potential.

The TR8 remains the rarest of the TR family, having been built in even smaller numbers than the TR5. The cars assembled at Canley and Solihull for production were all drophead models, but small numbers of fixed-head coupés were built at Speke in 1977, and these were exported to the USA for trials, eventually being sold off to the public.

Above
Decals on the front wings helped to distinguish the TR8 from its four-cylinder sisters.

Above and left
This particular Bordeaux Red TR8 spent the first 16 years of its life in storage, initially in the USA and then with a collector in the UK. In both cases, the car had been kept as an investment. It was eventually sold to Richard Connew, who had it converted to right-hand-drive and put it back on the road.

Right
Although the TR8 was sold for a time in the USA, no right-hand-drive home-market model was ever made available. This is a 1980 car, professionally converted from left-hand-drive and now owned by Triumph specialists Rimmer Bros. As the picture shows, it is not easy to distinguish a TR8 from a TR7 at first glance...

Facing page, bottom
...unless the badging is visible. The shaded decal badges for the TR8 were designed in the USA.

Facing page, top
The TR8's dashboard was broadly similar to that of the TR7. The neat three-spoke wheel was standard on US models and doubtless would have been standard on UK-market cars as well.

Left
The tan velour-faced upholstery was comparatively rare, but complements the metallic finish perfectly.

Right
Some cars were used for cosmetic development work at the factory. This one had colour-coded bumpers instead of the black units usually fitted to TR8s.

Above

The key to the whole thing, of course, was the 3.5-litre Rover V8 engine. It fitted neatly into the TR7 engine bay, which had been designed from the beginning to accommodate it.

Above left

This neat style of upholstery might have been standard on the UK-market cars.

Left

This TR8 fixed-head coupé is one of a batch built for evaluation in the USA, all with left-hand-drive, although this one was retained by BL in the UK for emissions testing. Since being sold in 1980, it has been maintained in original condition and still has its catalytic converters.

Above
A few right-hand-drive TR8s were built: some were used as development cars, while others were intended to form the press and demonstration fleet. This example, originally white, but since resprayed in metallic dark grey, is one of those cars.

Right
Characteristic of the TR8 were twin exhaust pipes. Note also the 'TR8' decal on the boot lid.

Above
Establishing what the UK-specification TR8s were intended to be like is problematical. This one, for example, wears both the TR8 decal and the wreath badge associated with TR7s built at the Rover plant in Solihull.

Left
To compensate for the scarcity of real TR8s, specialists in Britain began to offer professional V8 conversions of the TR7 during the 1980s. The engines used, as in this case, are generally Rover SD1 examples, although they are often tuned for additional performance.

127

Right

This TR7 in Persian Aqua was converted to V8 specification in 1991, using a new Rover V8 engine and as many genuine TR8 parts as possible. With the Holley carburettor and four-branch manifolds, its output is around 200bhp (149kW). Obviously, the brakes and suspension were also uprated.

TR8 SPECIFICATIONS

Engine	Eight-cylinder, OHV, carbureted (fuel-injected for California and for all 1981-season US models)
Capacity	3528cc (88.9mm bore, 71.1mm stroke)
Max. power	Carbureted models, 133bhp (99kW) DIN @ 5000rpm; fuel-injected models, 137bhp (102kW) DIN @ 5000rpm
Max. torque	Carbureted models, 174lb/ft (259kg/m) @ 3000rpm; fuel-injected models, 168lb/ft (250kg/m) @ 3250rpm
Transmission	Five-speed, manual all-synchromesh gearbox; optional three-speed automatic.
Suspension, front	Independent, with coil springs, MacPherson struts, anti-roll bar and telescopic dampers
Suspension, rear	Live axle, with coil springs, radius arms, anti-roll bar and telescopic dampers
Steering	Rack-and-pinion; power-assisted
Brakes	Discs on the front wheels and drums on the rear wheels; servo-assisted
Tyres	185/70x13 radial-ply
Length	13ft 8.5in (4.18m)
Width	5ft 6.2in (1.61m)
Height	4ft 1.9in (1.26m)
Wheelbase	7ft 1in (2.16m)
Max. speed	120mph (193km/h)
0-60mph (97km/h)	8.4sec
O'all fuel consumption	18.5mpg (7km/litre)
Production total	2715 (81 for home market*; 2634 for export)
	* Home-market cars were for evaluation only